Lecture Notes in Control and Information Sciences 378

Editors: M. Thoma, M. Morari

T0223114

Ricardo Femat, Gualberto Solis-Perales

Robust Synchronization of Chaotic Systems via Feedback

Springer

Authors

Ricardo Femat

IPICYT
Departamento de Matemáticas Aplicadas
Camino a la Presa San José 2055
Col. Lomas 4a. sección
C.P. 78216 San Luis Potosí
México
E-Mail: rfemat@ipicyt.edu.mx

Gualberto Solis-Perales

Universidad de Guadalajara
Centro Universitario de Ciencias Exactas e Ingenierías
División de Electrónica y Computación
Blvd. Marcelino García
Barragán 1421
Jal. C.P. 44430 Guadalajara
México
E-Mail: gualberto.solis@cucei.udg.mx

ISBN 978-3-540-69306-2 e-ISBN 978-3-540-69307-9

DOI 10.1007/978-3-540-69307-9

Lecture Notes in Control and Information Sciences ISSN 0170-8643

Library of Congress Control Number: 2008928642

Typeset & Cover Design: Scientific Publishing Services Pvt. Ltd., Chennai, India.

Printed in acid-free paper

5 4 3 2 1 0

springer.com

To you, beauty that has shared your time
and all my small conquests
along recent years.
Ricardo Femat

To Saidy, Regina, Estela and Antonio.
Gualberto Solís-Perales

Preface

This pages include the results derived during last ten years about both suppression and synchronization of chaotic -continuous time- systems. Along this time, our concept was to study how the intrinsic properties of dynamical systems can be exploited to suppress and to synchronize the chaotic behavior and what synchronization phenomena can be found under feedback interconnection. Our findings have caused surprise to us and have stimulated our astonishing capability. Perhaps, reader can imagine our faces with opens eyes like children seeing around objects; which are possibly obvious for others and novel for us. A compilation of our surprises about these findings is being described along this book. Book contains both objectives to share our amazement and to show our perspective on synchronization of chaotic systems. Thus, while we were writing the preface, we discussed its scope. Thinking as a book readers, we found that a preface should answer, in few words, the following question: What can the reader find in this book?, reader can find our steps toward understanding of chaotic behavior and the possibility of suppressing and synchronizing it. We firstly show the chaos suppression form experimental domain to potential implementation in high tech system as a levitation system based on High Temperature Superconductors (HTS). This chapter is used as departing point towards a more complicated problem the chaotic synchronization. Then, reader travels by the synchronization of the chaotic behavior world throughout distinct feedback approaches. An extension to classical feedback is first ste, where a classification of synchronization phenomena and potential implementation allow to go beyond the control application. Then, a small trip by geometrical tools to induce synchrony on chaos behavior leads us back to engineering. Thus, we approach landing onto time-discretization of feedback towards potential implementation in microcontrollers. Finally, extension on geometrical tools and synchronization are included in last chapter.

April 2008

Ricardo Femat
Gualberto Solís Perales

Contents

1 Introduction to Chaos Control: An Interdisciplinary Problem

1.1 Chaos Control Is Suppression or Synchronization

The foundational for chaos control problem is scientific as well as technological. In regard science, on the one hand, chaos control has two important contributions: (i) The controlled chaotic systems has allowed to understand that structured disorder and its entropy/information relationship extend the concept of determinism [1], [2] and (ii) departing from chaotification (inverse action of the chaos suppression) some questions have been opened on phenomena of the feedback dynamical systems [3]. Moreover, the chaos control impacts biomedical, life and engineering sciences; for example, it can be extended to control pathological rhythm in heart [4]. Now, regarding technological applications, the controlled chaotic systemsare important because of a desired frequency response can be induced. Nowadays, the scientific community has identified two problems in chaos control: suppression and synchronization. Among others, we can mention studies in physical devices (e.g., telescopes or lasers), biology/ecology (e.g., population dynamics or biodynamics) or biomedical systems (e.g., heart rhythm or brain activity). Thus, for instance, controlled current-modulation can be entered as excitation from a nonlinear circuit into semiconductors lasers by feeding back the laser frequency response (see Figure 1 in [5]). Henceforth, scientific community has taken possession of the challenge of exploring control techniques such that (i) a family of driving force can command classes of chaotic systems [6], (ii) the synthesis of mathematical expressions for the control force accounts the frequency response [7], and (iii) energy requirements by the control force are accounted (for example to avoid saturation or deterioration in control devices) [8]. In addition, the mathematical models of the driving force is desired to be simple and easy to implement experimentally. A simple form is the linear models of driving forces; which can be expressed in the frequency (Laplace) or time domain and they have been already used to suppress chaotic behavior [7], [9].

In grosso, the chaos suppression problem can be defined as the stabilization of unstable periodic orbits (UPO's) of a chaotic attractor in equilibrium points or periodic orbits with period n embedded into the chaotic attractor [10]. Since the

R. Femat & G. Solis-Perales: Robust Syn. of Chaotic Sys. Via Feedback, LNCIS 378, pp. 1–5, 2008.
springerlink.com

seminal paper by Ott,Grebogi and Yorke [11] was published, several control schemes have been proposed to suppress chaos. Continuous- and discrete-time approaches can be found in open literature (see, for example, [12] and [13]). Some feedback controllers have been designed from robustness against noisy environment [14]. Others have been proposed as robust approaches for state feedback control [15] and few schemes have been designed in frequency domain. In this sense, integral actions have shown capability to stabilize chaotic systems in equilibrium points and periodic orbits [7], [9]. Nevertheless, the control cost is oftenomittedin reports of chaos suppression. Thus, the following question rises: can we design a feedback control (driving force) with robustness and optimality issues for chaos suppression ?. The problem is not an easy task if we consider that: (i) Controversy on robust, optimal and fragility issues is open in control theory, (ii) Chaotic systems are, by nature, highly sensitive to initial conditions and parametric variationsand (iii) The chaotic systems are nonlinear with continuous spectrum in frequency which can complicate the synthesis of the frequency domain driving forces.

Synchronization of chaotic systems is an interesting topic that, since early 90's, has caught the attention of the nonlinear science community. Two research directions have been already conformed in synchronizing chaos: (i) analysis and (ii) synthesis. Analysis problem comprises (a) the classification of synchronization phenomena [16], [17]; (b) the comprehension of the synchronization properties as, for instance, robustness [18] or geometry [19], [20]; and (c) the construction of a general framework for unifying chaotic synchronization [17], [21]. On the other hand, synthesis of synchronization systems concerns the problem of finding the control force such that two chaotic systems share time evolution in some sense. Both analysis and synthesis directions are active research areas and one of the current challenges is to achieve and explain synchronization of chaotic system with different model. In fact, the study of the chaotic synchronization with different models makes sense in several systems (see references within [22], [23], [24] and [25]). Among others, we can account those with different fractal dimension [22], neural levels [23],[24], message transmission [25] or respiratory/ circulatory coupling [24].

Inregardto analysis in strictly-different systems, the studies have been focussedonthe existence of synchronization manifolds for coupled systems and such manifolds strongly depend on measures from Lyapunov exponents [19], [20]. Synchronization of different models has been analysed in nonidentical space-extended systems (for the case of parameter mismatching)[26] and structurally nonequivalent system including delay [22]. In [19] chaotic synchronization has been also analysed from invariant manifolds in terms of the existence of a diffeomorphism between the attractor of the coupled systems; which is closely related to generalized synchronization (GS). Josic [19] had included synchronization of different systems, and illustrative examples show the existence of synchronization manifolds; e.g, between Rössler and Lorenz. This analysis departs from rigorous definitions, and is deep for the complete synchronization (i.e., the synchronization of *all* master states with *all* corresponding states of the slave system, [16]). Unfortunately, such a formalism for other synchronization phenomena (as, for example, the partial-state synchronization [16]) is still obscure.

Concerning the synthesis, by the end of 90's [27], some efforts have been done to synchronize chaotic systems with different model. The underlying idea is to find a synchronization force such that the existence of a synchronization manifold can be assured. In this manner, several synchronization phenomena can be found [16], and different design techniques have been exploited [23], [24], [25], [26], [27], [28] and [29]. A sliding-mode feedback scheme was proposed in [28], where, following ideas in [27], the synchronization of Duffing (master) and van der Pol (slave) was performed. A feedback scheme based on active control was presented in [29]. Moreover, open-loop schemes have been proposed as well. In [23] a control force is *trained* for inducing synchronization in discrete-time chaotic systems. Although robustness can be lost in open-loop control systems, Xiaofeng-Lai's results [23] are interesting because of they show that chaotic synchronization can be achieved even under open-loop interconnection. More recently, synchronization of systems with different order has been reported [24]. A nonlinear feedback interconnection under lack knowledge was performed. Thus, the synchronization has been attained between Duffing equation and the canonical-plane projection of the Chua's circuit. Synchronizing force for chaotic systems in the triangular form has been recently synthesized from observer-based sliding-mode schemes by Feki [30] for similar models (Lur'e systems) and by Yang [28] for driven strictly-different systems (Duffing-van der Pol). This is an advantage because of such a triangular form can be physically realized [31], [32] in order to experimentally corroborate the synchronization schemes and can be derived from Lie derivatives in dynamical systems [17], [33].

In the following chapters the problem of controlling chaos, *i.e.*, suppression and synchronization, is dealt from the perspective of control theory but conserving the scope of nonlinear science. The idea is to discuss inductively this topic from simple approaches to more complicated structures. Exercises are proposed in order to lead the reader into subsequent sections and chapters. Few entire sections are devoted to examples to illustrate potential applications. Even experimental implementations are developed to show some practical aspects of the proposals. Thus, the chaos suppression is dealt in second chapter departing from Laplace controllers to derive an output feedback controller as preliminary step. The chapter includes second order driven systems; which allows to introduce the notion for disturbance attenuation (*i.e.*, driven force compensation). The application of chaos suppression on a magnet-superconductor levitation close the chapter. Then, in Chapter 3, the chaotic synchronization is addressed via linear feedback control. The proposal departs from Laplace controllers in second -order driven systems and, passing through some phenomena description, a framework is derived for the synchronization via proportional-integral control. An experimental application to secure communication is included, and a robust analysis is sketched. The Chapter 4 is the most important. Ageneral framework is discussed for finite n-dimensional chaotic systems via robust asymptotic feedback. The secure communication is used as example of physical implementation. Finally, Chapter 5 has been written in searching a general theory on chaotic synchronization. The underlying idea is to open the *synchronizability* notion. That is, the property of the vector fields such that is possible to synchronize, in some sense related with any of the diverse reported phenomena, nonlinear dynamical systems. Lie algebras of vector fields is exploited, inalgorithmic sense, to discuss the point. Complete-state practical and generalized synchronization are included in analysis.

References

[1] Hayles, N.K.: Chaos bound. Orderly disorder in conterporary literature and science. Cornel Univ. Press, USA (1990)

[2] Bricmont, J.: Science of Chaos or chaos of science? In: Gross, P.R., Levitt, N., Lewis, M.W. (eds.) The flight from science and reason, USA. Annals of the New York Academy of Sciences, vol. 775, pp. 131–175 (1998)

[3] Lu, J., Yu, X., Chen, G.: Generating chaotic atractors with multiple merged basins of atractions: A switching piecewise-linear control approach. IEEE Trans. Circ. Syst. I 50, 198–207 (2003)

[4] Christini, D.J., Collins, J.J.: Using chaos control and tracking to suppres a pathological nonchaotic rhythm in a cardiac model. Phys. Rev. E 53, 49–52 (1996)

[5] Vasilév, P.P., White, I.H., Gowar, J.: Fast phenomena in semiconductor lassers. Rep. Prog. Phys. 63, 1997–2042 (2000)

[6] Booker, S.M.: A family of optimal excitation for inducing complex dynamics in planar dynamical systems. Nonlinearity 13, 145–163 (2000)

[7] Femat, R., Capistrán-Tobias, D., Solís-Perales, G.: Laplace domain controlers for chaos control. Phys. Letts A. 252, 27–36 (1999)

[8] Sarasola, C., Torrealdea, F.J., d'Anjou, A., Graña, M.: Cost of synchronizing different chaotic systems. Math. Comp. Simulation 58, 309–327 (2002)

[9] Puebla, H., Alvarez-Ramirez, J., Cervantes, I.: A simple tracking control for Chuaś circuit. IEEE Trans. Circ. and Syt. I 50, 280–284 (2003)

[10] Aguirre, L.A., Billings, S.A.: Closed-loop suppresion of chaos in nonlinear driven oscillators. J. Nonlinear Sci. 5, 189–206 (1995)

[11] Ott, E., Grebogi, C., Yorke, J.A.: Controling chaos. Phys. Rev. Letts 64, 1196–1199 (1990)

[12] Alvarez-Ramirez, J., Garrido, R., Femat, R.: Control of systems with friction. Phys. Rev. E 51, 6235–6238 (1995)

[13] Pyragas, K.: Continuous control of chaos by self-controling feedback. Phys. Letts. A 170, 421–428 (1992)

[14] Cazelles, B., Boujdjema, G., Chau, N.P.: Adaptive control of systems in a noisy environment. Phys. Lett. A 196, 326–330 (1995)

[15] Alvarez-Ramirez, J., Femat, R., Gonzalez, J.: A time delay coordinates strategy to control a class of chaotic oscillators. Phys. Letts. A 211, 41–45 (1996)

[16] Femat, R., Solis-Perales, G.: On the chaos synchronization phenomena. Phys. Letts. A 262, 50 (1999)

[17] Brown, R., Kocarev, L.: An unifying definition of synchronization for dynamical systems. Chaos 10, 344 (2000)

[18] Kocarev, L., Parlitz, U., Brown, R.: Robust synchronization of chaotic systems. Phys. Rev. E 61, 3716 (2000)

[19] Josic, K.: Synchronization of chaotic systems and invariant manifolds. Nonlinearity 13, 1321 (2000)

[20] Martens, M., Pécou, E., Tresser, C., Worfolk, P.: On the geometry of master-slave synchronization. Chaos 12, 316 (2002)

[21] Boccaleti, S., Pecora, L.M., Pelaez, A.: Unifying framework for synchronization of coupled dynamical systems. Phys. Rev. E 63, 066219-1 (2001)

[22] Boccaleti, S., Valladares, D.L., Kurths, J., Maza, D., Mancini, H.: Synchronization of chaotic structuraly nonequivalent systems. Phys. Rev. E 61, 3712 (2000)

[23] Xiaofeng, G., Lai, C.H.: On synchronization of different chaotic oscillators. Chaos, Solitons and Fractals 11, 1231 (2000)
[24] Femat, R., Solís-Perales, G.: Synchronization of chaotic systems with different order. Phys. Rev. E 65, 036226-1 (2002)
[25] Femat, R., Jauregui-Ortiz, R., Solís-Perales, G.: A chaos-based communicaiton scheme via robust asymptotic feedback. IEEE Circ. Syst. I 48, 1161 (2001)
[26] Boccaletti, S., Bragard, J., Arecchi, F.T., Mancini, H.: Synchronization in nonidentical extended systems. Phys. Rev. Letts. 83, 539 (1999)
[27] Femat, R., Alvarez-Ramírez, J.: Synchronization of two strictly different chaotic oscillators. Phys. Letts. A 236 (1997)
[28] Yang, T., Shao, H.H.: Synchronizing chaotic dynamics with uncertainties based on sliding mode control design. Phys. Rev. E 65, 046210-1 (2002)
[29] Ho, M.C., Hung, Y.C.: Synchronization of two different systems by using generalized active control. Phys. Letts. A 301, 424 (2002)
[30] Feki, M.: Observer-based exact synchronization of ideal and mismatched chaotic systems. Phys. Letts. A 309, 53 (2003)
[31] Sprott, J.C.: A new class of chaotic circuits. Phys. Letts. A 266, 19 (2000)
[32] Malasoma, J.M.: A new class of minimal chaotic flows. Phys. Letts. A 305, 52 (2002)
[33] Femat, R.: An extension to chaos control via Lie derivatives: Fully linearizable systems. Chaos 12, 1207 (2002)

2 Chaos Suppression with Least Prior Knowledge: Continuous Time Feedback

2.1 Experimental Space in Frequency Domain

Firstly, we present the frequency spectrum as an alternative procedure for studying feedback effects onto chaotic systems. It is well known that continuous power spectrum is an important feature of chaotic systems. This fact can be used for distinguishing a chaotic system from time series. Although power spectrum is not definitive for identifying chaotic systems [1], power spectrum allows us to understand the effect of feedback onto chaotic systems in terms of the control parameters. In some sense, power spectrum can be seem as a dynamic bifurcation diagram [2], in fact, to study bifurcation of chaotic systems can be an important tool. To this end, there are two basic concepts: (i) Dynamics of a given nonlinear system can be approached by $\dot{x} = f(x;\pi)$, where f: $\mathbb{R}^n \to \mathbb{R}^n$ and $\pi \in \mathbb{R}^p$ is a set of parameters (which can be a constant or time functions). Thus, qualitative changes of the system behavior can be induced for certain values of the parameters $\pi \in \mathbb{R}^p$, i.e., it is possible that the nonlinear system displays chaos. (ii) such qualitative changes in dynamics of a nonlinear system can be observed due to parametric variations (i.e., bifurcation diagram). Hence, bifurcation diagrams are very important and can be also used as characterization procedure for chaotic systems.

In this section, the power spectrum density (PSD) is presented as an option to obtain a bifurcation diagram from time series of the controlled chaotic system. Such a procedure results in a 3-D diagram in such manner that the behavior of the nonlinear system can be studied in the same sense than bifurcation. The 3-D PSD diagram is constructed by plotting amplitude versus frequency and control gain. 3-D PSD has the following advantage, it is capable to account dynamic behavior from parametric variation or forcing. This is, since the chaotic behavior can be yielded from (a) varying parameters values or (b) exciting the system from external signal, e.g., exciting force or noise (see for instance [3],[4]) a procedure is desired to understand the PSD of chaos in similar manner than bifurcation. Lorenz system has been chosen to illustrate the study from 3-D PSD, which is controlled via PI feedback. Such a controller

R. Femat & G. Solis-Perales: Robust Syn. of Chaotic Sys. Via Feedback, LNCIS 378, pp. 7–50, 2008.
springerlink.com

was chosen due to it is able to lead the chaotic system around the reference point (which can be prescribed, without lost of generality, as the origin). Moreover, PI controller is a simple and classical feedback. That is, this section deals with the effect of a given feedback controller onto chaotic systems.

The goal is to study closed-loop chaotic systems. Thus, it is convenient to discuss the change that a nonlinear system suffers under feedback interconnection. Figure 2.1 shows the simplified block diagram of both open-loop and closed-loop systems. The open-loop corresponds to the entering of initial conditions into the nonlinear system. Whereas frequency ω and amplitude α enter into exciting force (see Fig. 2.1a). In this way, the dynamical behaviour of the open-loop system depends on the exciting force input ω and α while time evolution depends on initial conditions $x_0 = x(t = 0)$. In principle, the chaotic behaviour can be controlled from suitable values of the frequency ω and amplitude α, which are entering into the exciting force system. Moreover, chaotic behaviour can be induced for a proper values of the frequency ω and amplitude α. This kind of interconnection (open-loop) has been studied by several authors (for instance see [1], [5], [6] and [7]). On the other hand, a closed-loop system consists in the same open-loop system and an additional block, which connects the feedback. Same inputs enters into nonlinear system and exciting force. In addition, the output of the nonlinear system (which can be represented by the time series) is entering to the controller from the feedback loop. Such an output is often referenced as the observable or measured state. This kind of interconnection has been widely studied in last decade (see for example, [8], [9], [10], [11]). Closed-loop scheme for chaos control has some advantages, which are discussed in [12].

Now, let us consider the driven oscillators, whose model becomes

$$\dot{x} = f(x;\pi) + \tau_e(t) \tag{2.1}$$

where $x \in \mathbb{R}^n, \pi \in \mathbb{R}^P$ stand for states and parameters, $f: \mathbb{R}^n \to \mathbb{R}^n$ is a nonlinear and smooth function and $\tau_e(t)$ is the exciting force. System (2.1) is the mathematical representation of the diagram block in Fig. 2.1a. It should be pointed out that the

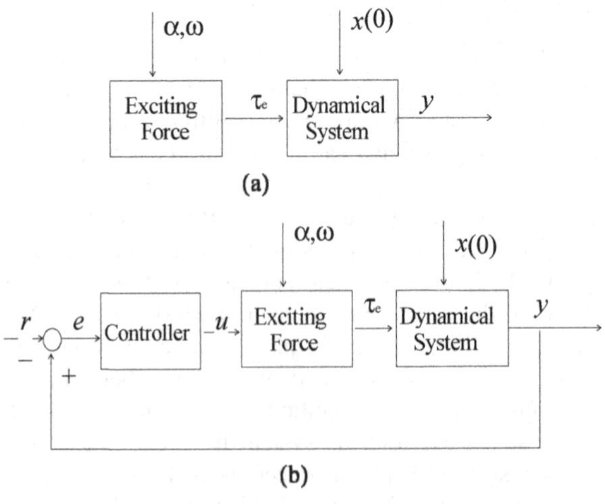

Fig. 2.1. (a) Open and (b) closed-loop of the controlled chaotic systems

exciting force can be yielded by noise or periodic signals. Thus, if system (2.1) represents a second-order driven oscillator, then, according to Poincaré-Bendixon theorem [1], the periodic exciting force is required to induce chaos into system (2.1). On the contrary, for autonomous dynamical systems whose order is larger than two, the effect of the exciting force can induce or suppress the chaotic behaviour in system (2.1). In fact, the effect of some kinds of external signals (as noise) have been studied by Anishenko [6], who used some methods of stochastic theory to compute bifurcations of autonomous systems. Stochastic methods make sense for studying bifurcations of dynamical systems in presence of noise. However, if exciting force is deterministic, as in second-order driven oscillators, a more simple procedure is desired.

This class of system is topologically and geometrically interesting [1] and formal analysis has been reported in the closed-loop control of second-order driven systems [11]. The importance of system (2.1) raises from the fact that diverse physical systems can be modelled with equation (2.1). For example, if $n = 2$, friction systems as telescopes or sliding sticks [53]. Also the van der Pol equation is included in the form (2.1), and is widely used as model in performance of radio tubes [16], biological systems [45] or electrical circuits. The system (2.1) is a simple nonlinear system that has been used as a benchmark for the study of chaos suppression [12],[13] and synchronization [32]. Indeed, three dimensional non-autonomous systems can be represented by (2.1). As for instance, the magnetic levitation of superconductor Type II is modelled by (2.1) and its closed-loop control has been was recently reported [42]. Also the open-loop control can be used to suppress chaos in 3-dimensional non-autonomous systems. Open-loop control can be performed by choosing parameters of: (i) the system or (ii) the exciting force. The following exercise illustrates the open-loop control in 3-dimensional systems. Nevertheless, notice that open-loop control does not involve measurements of the actual state of the system; which is a drawback because corrections cannot be induced.

Exercise 2.1 *Open-loop control in 3-dimensional non-autonomous systems.* Consider the 3-dimensional non-autonomous system $\dot{x}_1 = x_2$; $\dot{x}_2 = 0.125(x_1^2 + 3x_3^2)x_1 - kx_2 + \alpha\cos(t)$; $\dot{x}_3 = 0.125k(x_3^2 + 3x_1^2)x_3 + \gamma$. Assume that $\gamma = 0.03$ and $k = 0.5$ hold constant. Plot the attractor of the system for values of α in the interval [0.2,0.3]. Note that the variation of the amplitude value is related to the open-loop chaos suppression. Moreover, notice that the system is highly sensitive to small variations in amplitude. That is, for $\alpha = 0.22$ the systems is periodic whereas for $\alpha = 0.25$ the system displays chaotic behaviour [6].

On the other hand, if a scalar output (i.e., one measured state or its time series) of the system dynamics in the closed-loop system is given by returning the information from the measurement $y \in \mathbb{R}$. Thus the closed-loop system can be written as

$$\begin{aligned} \dot{x} &= f(x;\pi) + \tau_e(t) + g(x)u \\ \dot{e} &= r - y \end{aligned} \tag{2.2}$$

where $e = r - y$ is the control error, r is the prescribed reference (which can be chosen, without lost of generality, as the origin), u denotes the feedback control force and is given $u = k_c(r - y) + k_c\tau_I^{-1}\int(r - y)d\sigma$ (PI controller), $g(x)$ is a vector field, which determines the control channel, and k_c denotes the feedback gain. Note that the chaotic oscillator (2.1) is modified under feedback. One can prove that there exists a value of the feedback gain k_c such that the system (2.1) is stable at the reference point [13]. We are interested to study the dynamics of system (2.2), i.e., the chaotic oscillator under feedback control. Indeed, the feedback gain k_c is presented as a bifurcation parameter of the controlled system (2.2). In particular, the bifurcation of the controlled system (2.2) is discussed in this section via FFT and power spectrum. By extension, the bifurcation diagram serves to explain how the orbits of a nonlinear dynamical system change [1]. We belief that FFT and power spectrum are effective tools for comprehension of the chaos control [14] in same sense than bifurcation diagrams, i.e., to explain the change in orbits of a given nonlinear system from a measured state or time series.

Let us consider Lorenz equation as an illustrative example, whose open-loop model is given by

$$\begin{aligned} \dot{x}_1 &= \pi_1(x_2 - x_1) \\ \dot{x}_2 &= \pi_2 x_1 - x_2 - x_1 x_3 \\ \dot{x}_3 &= -\pi_3 x_3 + x_1 x_2 \end{aligned} \tag{2.3}$$

where $\pi \in \mathbb{R}^3$ is a set of parameters of the system. Although, Lorenz equation was obtained in 1963 from the mathematical model of convective systems [15], it has been used for modelling several dynamical systems (for example multi-modal laser, see Chapter 10 in [16]). System (2.3) can be controlled via parameter modulation, i.e., open-loop interconnection by choosing the system parameters. To this end a parameter is varied without feedback of any measured state. Figure 2.2 shows the phase portrait of the Lorenz equation for two different parameters. Same initial conditions were arbitrarily chosen whereas two different values were selected only for π_2. Note that, for a given value of the system parameter, the system (2.3) behaves chaotically while for another value the trajectories of the same system converge to an equilibrium point. Of course, convergence of a given equilibrium point depends on initial conditions. Indeed, existence of multiple equilibrium points adds an additional complexity degree for bifurcation studies of nonlinear systems [17]. However, we are not interested to discuss this topic. The interested reader is encouraged to read the seminal books [1] and [17]. Here, we are interested in discussing the effect of the loop interaction into (or onto) nonlinear chaotic systems. In what follows we show the closed-loop control of the system (2.3).

Exercise 2.2. By taking the system in Exercise 2.1. Plot the power spectrum density (PSD) versus the amplitude of the exciting force α in the interval [0.2,0.3]. Note that open-loop control via parameters of the exciting force is analogous to the open-loop control via system parameters. In fact, a 3-dimension plot of the PSD can be also computed by taking the parameters k and/or γ. Are there regions where the system exhibits many fundamental frequencies and others where the system has no oscillatory behaviour?

Thus, from previous digression, one can expect that PSD of the Lorenz equation varies with the parameter values. Perhaps, the effect of the parameter modulation is clearly seen in second-order driven systems. In such a case, the open-loop control can be performed via the amplitude and/or frequency of the exciting force. This is, the

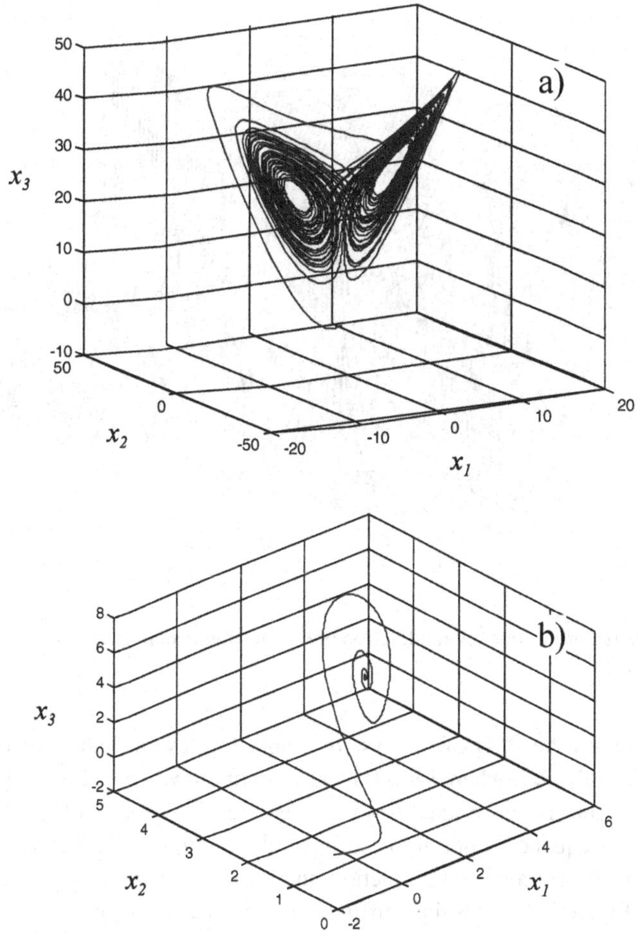

Fig. 2.2. Dynamic behavior of the Lorenz system under parameter modulation, open-lop interconection. a) $\pi = (10,28,8/3)$ and b) $\pi = (10,5,8/3)$.

open-loop interconnection of the driven oscillators results in the following mathematical model: $\dot{x}_1 = x_2,\ \dot{x}_2 = \varphi(x;\pi) + \tau_e(t)$, where $\varphi(x)$ is a nonlinear function (which defines the mechanical, electrical or physical system), $\pi \in \mathbb{R}^p$ is a set of system parameter (which can be time function) and $\tau_e(t)$ denotes the exciting force (which is often chosen smooth and bounded). We can consider that second-order driven system is given by the driven Duffing equation [1]. Therefore, $\varphi(x;\pi) = \pi_1 x_1 - x_1^3 - \pi_2 x_2$, where π_1 and π_2 are constant and $\tau_e(t) = \alpha\cos(\omega_e t)$, where the constant α denotes the amplitude and ω_e is the frequency. If we choose $\pi = (1,0.15)$ and $\alpha = 0.275$, one can induce frequency modulation by means of the variation in ω_e. We have chosen the position as the system output, i.e., $y = x_1$.

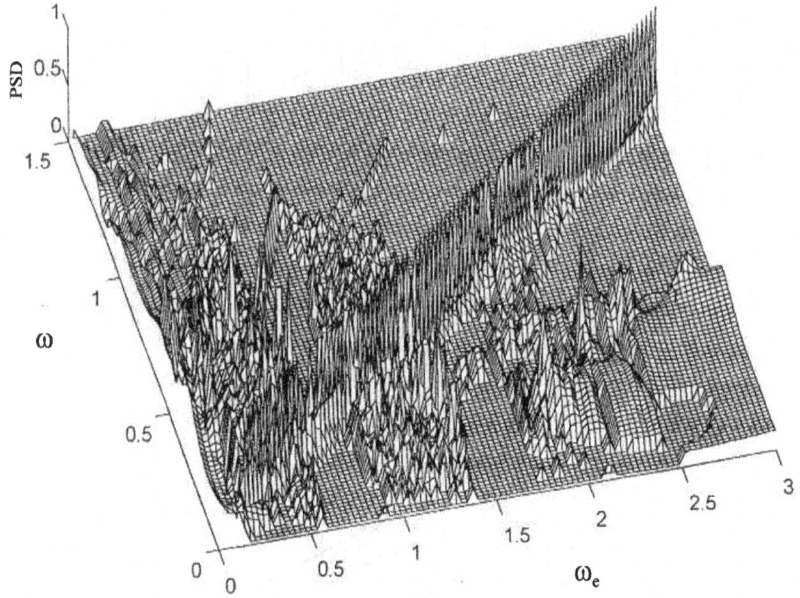

Fig. 2.3. PSD vs. (ω, ω_e) frequency modulation of the Duffing equation in an open-loop control scheme

Figure 2.3 shows the PSD of the system output for several values of the exciting frequency $\omega_e \in [\underline{\omega}_e,\bar{\omega}_e]$, where $\underline{\omega}_e \geq 0$ means minimum value and $\bar{\omega}_e < \infty$ maximum value. Such a picture is constructed from Fast Fourier Transform (FFT) of the output system for each frequency value into the interval $[\underline{\omega}_e,\bar{\omega}_e]$. Note that in spite of the exciting force is always a periodic function, the response of the Duffing equation can display chaos. Figure 2.4 shows the Duffing equation phase portrait for same values of the system parameters than Figure 2.3 and initial conditions $x(0) = (-0.5,1)^T$. In Figure 2.4a, Duffing equation is not driven whereas in Figure 2.4b the driven frequency $\omega = 1.1$. The parameter modulation of chaotic systems has been used for controlling chaos and several applications have been reported (e.g., secure communication [18]).

On the other hand, closed-loop system can be also understood in similar manner than open-loop interconnection. In the sense that we can choose the parameters of the feedback control an compute the behaviour. Let us choose two classical examples: Lorenz and Duffing equation. Both Lorenz and Duffing equations have been also used to study chaos control via nonlinear feedback [12], [1]. Indeed, authors in [13] proved that chaos in system (2.3) can be controlled via PI feedback. The goal of the 3-D PSD is to show the interaction yielded by the feedback parameter k_c, in other words, to illustrate the bifurcation effect given by the control parameter. Now, let us suppose that the vector field $g(x) = [1,0,0]^T$ and that $y = x_1$. Then the closed-loop of the Lorenz equation becomes

$$\begin{aligned}
\dot{x}_1 &= \pi_1 (x_2 - x_1) + u \\
\dot{x}_2 &= \pi_2 x_1 - x_2 - x_1 x_3 \\
\dot{x}_3 &= -\pi_3 x_3 + x_1 x_2 \\
\dot{x}_4 &= x_1 - r
\end{aligned} \tag{2.4}$$

where the prescribed reference is arbitrarily chosen $r = 0$ (which means that we are interested to control system (2.4) around origin). Since by definition $e(t) = x_1(t) - r(t)$ and $r(t) = 0$ for all time $t \geq 0$ and $x_4 = \int e(\sigma)d\sigma$, the PI feedback is given by $u = k_c x_1 + k_c \int x_1(\sigma)d\sigma$. Now, if the parameter values of the system are chosen such that the open-loop system is chaotic (i.e, $\pi = (\pi_1,\pi_2,\pi_3) = (10,28,8/3)$), then the unique bifurcation parameter is the feedback gain k_c. Figure 2.5 shows the 3-D PSD for the closed-loop system (2.4).

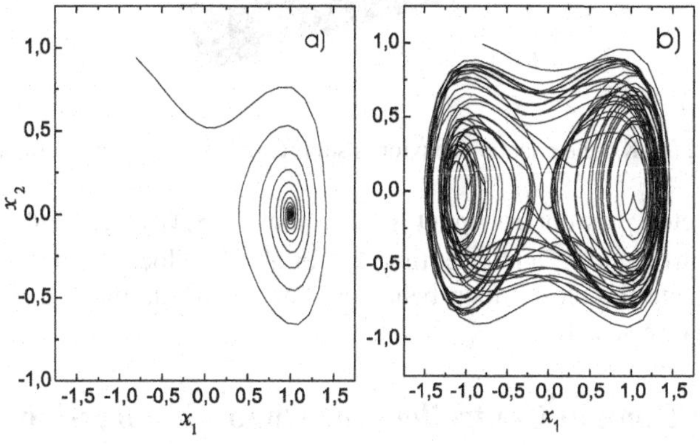

Fig. 2.4. Phase portrait of the Duffing system (a) for the driving force $\tau(t) = 0$ and (b) the driving force $\tau(t) \neq 0$

Note that the picture allows to identify the regions where system (2.4) displays different behavior. Indeed, if feedback gain k_c varies, then different attractors can be obtained from the closed-loop (2.4). In the same way, if position is the output of the second-order driven system and $g(x) = (1,0)^T$, the closed-loop becomes $\dot{x}_1 = x_2 + u$, $\dot{x}_2 = \varphi(x;\pi) + \tau_e(t)$, where u is given by the PI controller, $\varphi(x;\pi) = \pi_1 x_1 - x_1^3 - \pi_2 x_2$ and $\tau_e(t) = \alpha\cos(\omega_e t)$.

Figure 2.6 shows the PSD versus (k_c,ω), where ω is the frequency computed in FFT. The values of the system parameters were chosen in such a way that open-loop Duffing oscillator is chaotic, i.e., $\pi = (1,0.15)$, $\alpha = 0.275$ and $\omega_e = 1.1$. The chaos control in frequency domain and stability of the closed-loop is discussed in following sections. A detailed discussion of the PI controller is in next section. In fact, such as we will see in Chapter 3, several effects of the PI feedback can be found. These effects result in diverse synchronization phenomena.

Exercise 2.3. *PSD form the Proportional-Integral controller for chaos suppression in 3-D non-autonomous systems.* Consider the system in Exercise 2.1. Is there a feedback interconnection such that a PI control attains the chaos suppression?. Hint: Try

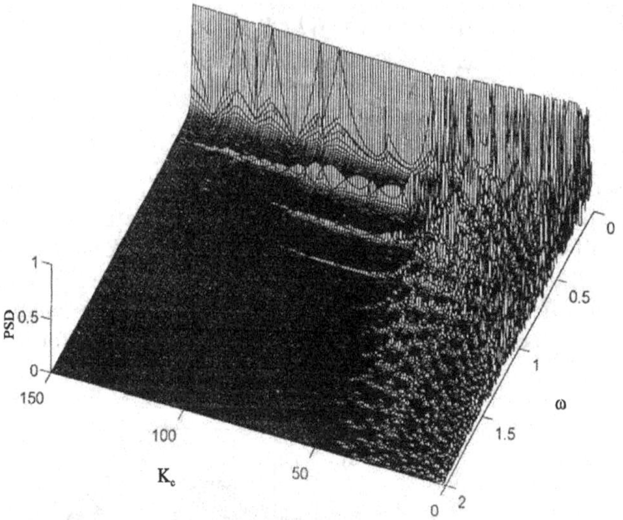

Fig. 2.5. PSD vs. (ω, K_c) the chaotic behavior is suppressed as the control parameter increases

with the parameter γ as control input $\gamma = K_c(x_3 - x_3^*) + K_c\tau_I^{-1}\int(x_3 - x_3^*)d\sigma$; i.e., find the value of the control gain K_c and reset time τ_I^{-1} for absolute local stability at $x_3^* = 0$. We suggest to simulate the closed-loop behaviour for a α in the interval [0.2,0.3] and k constant as in Exercise 2.1.

2.2 Laplace Domain Controllers for Chaos Suppression

Several feedback strategies has been proposed to focussed the problem of chaos suppression. Robust asymptotic control [20],[21],[22], Lyapunov-based methods [11],

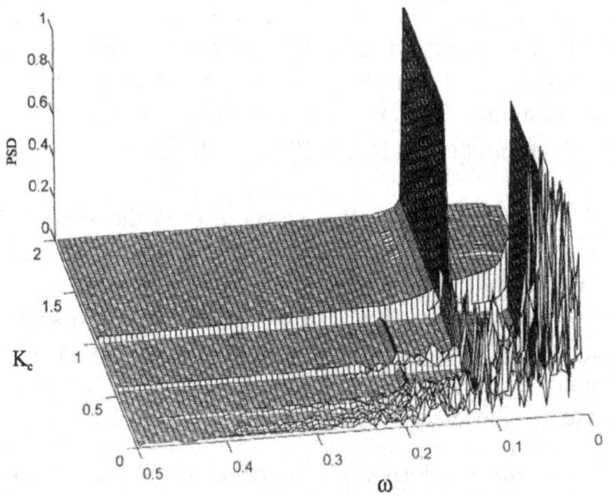

Fig. 2.6. PSD vs. (ω, k_{ce}) chaos suppression in the Duffing oscillator as the control parameter increases

time domain coordinates [23], linear adaptive schemes and nonlinear adaptive strategies [24],[25] and invariant manifold - based constructions [26] have been developed. Close-loop schemes are more robust [19]. Robustness is an important property of chaos control [27] due to chaotic systems can be highly sensitive to initial conditions and parameters uncertainties. As matter of fact, parameter uncertainties can be accounted from adaptive schemes, and PI-like controllers can be synthetised without knowledge about system parameters.

The linear and nonlinear strategies of the adaptive control theory allow the chaos control in spite of the parameters values of the system were unknown or they are time varying [21],[22]. Most of the adaptive control schemes are based on the dynamic parameter estimation. The estimated parameters are used into a reference model in such way that the dynamical behavior of the reference model is "adapted" to the real system. However, these adaptive feedback schemes have a disadvantage: *The structure of the model parameter must be known* [28]. Although adaptive control can be applied to several problem of the chaos control (as synchronization, suppression and stabilization), such application is restricted due to the parameter requirement (for example, it has not been shown that two strictly different oscillators can be synchronized by means of this kind of adaptive feedback). Nevertheless, a novel adaptive scheme is derived by lumping the uncertain terms in such manner that it is not necessary to estimate each parameter. Thus structure of model parameters is not needed. In following sections, this idea and its rationale are discussed.

Ott, Grebogi and Yorke (OGY) [29] have suggested a strategy to stabilize periodic orbits embedded in a chaotic attractor. The OGY method makes use of the on-line construction of local invariant manifolds of the target orbit to derive a controller which counteracts the unstable directions of the orbit. Although experimental applications of OGY strategy have been carried out [30], the method has some limitations. For example, it can stabilize only those periodic orbits whose maximal Lyapunov exponent is smaller than the reciprocal of the time interval between parameter changes. Since the parameter variations are small, the fluctuation noise can induce bursts (the bursts are more frequent for large noise) of the system into the region far of the target orbit [29]. The OGÝs control scheme has feedback structure [29]. The main idea is to construct an invariant manifold of the target orbit. In this way, the controller counteracts the unstable directions of unstable periodic orbits. However, because of its sensitivity to parameters and noise, this class of controller is not robust against uncertainties. In consequence the performance of the OGÝs scheme could be not acceptable. Some modifications to the original OGÝs strategy provided certain robustness margin. Such modifications are mainly based on adaptive control schemes [31]. Nevertheless, adaptive schemes have one more drawback: *The order of the controller increases as the number of time-varying parameters of the system.* Thus, adaptive based feedback schemes can improve performance of OGY low-order dynamical systems [31], but if the high-order system has many parameters the order of the controller will increase [28]. Moreover, if the system is non-linearly parameterized the physical implementation of the feedback control is quite difficult [28].

The above mentioned design algorithms depart from the internal model of the system to be controlled. For instance, although the robust asymptotic controller does not require *a priori* information about the system model, this design algorithm is based on an internal model procedure. This is, by departing from the system structure like second order driven oscillators, a robust asymptotic controller can be obtained. Such feedback controller yields chaos suppression [20]. Nevertheless, from the control systems viewpoint, it can be desirable a control law that leads to chaos control with least prior knowledge about the internal model of the dynamical system with robust properties. Here we discuss the robustness of the PI-controller and its relation with the robust asymptotic stability. The robust asymptotic controller is designed by means of the following procedure: (i) The uncertainties are lumped in a nonlinear function, (ii) the lumping nonlinear function is interpreted as an augmented state in such way that the extended system is dynamically equivalent to the original one, (iii) in order to obtain an estimate of the augmented state, a state estimator is designed for the extended system and (iv) the estimated value of the uncertainties is provided to the control law (via the estimated value of the augmented state). The above procedure yields a feedback controller which leads to chaos control in spite of modeling errors, parametric variations and/or external perturbations [20]. In fact, the robust asymptotic controller is able to perform the synchronization between two strictly different second order-driven oscillators [32].

In this section, two continuous-time feedback controllers are studied. The first one is the PI-controller. As we shall see, the PI-controller is able to stabilize chaotic systems around prescribed points. If the control goal is to track a prescribed trajectory, then the PI-controller does not achieve the chaos control goal. The second one, is a modified PI-controller. The modified PI-controller has the following three parts: (i) Proportional, (ii) Integral and (iii) Quadratic integral. The first two parts can be found in classical control theory literature [33],[34]. The last one has been recently developed and it has been called PII^2 (Proportional-Integral-Quadratic-Integral) [35]. It has been shown that PII^2 controller is a classical Proportional-Integral controller (PI) with enhanced uncertainty estimation capabilities. Physically, the idea behind the controller structure is that a dynamic compensator detects uncertainties (external perturbations, modelling errors and/or parametric variations) and take control actions to suppress and stabilize chaos. In this sense, the PII^2 controller departs from the same basic idea that the controller reported in [6]. In consequence, the PII^2 controller is robust against the uncertainties above mentioned. However, the procedure to design the PII^2 controller does not depart from model of the dynamical system. In fact, the design procedure of the PII^2 feedback departs from geometrical control theory [36],[37]. The design algorithm consists in coordinate transformations, which are based on Lie derivatives. The procedure is sketched below and is detailed in next section.

The main idea behind the control theory is to lead the system output into a desired reference in spite of lack knowledge about the dynamical systems that yields the output. The output signal, y, is a natural trajectory of the dynamical system P. This is, without control actions (open-loop), the system yields the observable y. The objective of the control block is to lead the output signal into a desired signal (namely reference, r is constant in regulation and a time-function for tracking) by means of a control command, u. To this end, the controller requires the actual value of the system output; in other words the feedback signal is the observable, y. Hence, in order to make its work, the controller needs to know at least the difference between the reference and the output signal, $e = y - r$. (See Fig. 2.1b).

Laplace domain characterization offers a very simple and elegant methodology which results in input-output models from frequency response. Such input-output models are widely used for control purposes [33],[34]. Input-output models provide straightforward analysis of how dynamical system to be controlled reacts to several external influences. There are established criteria to study the sensitivity of the system output respect to external disturbances (robust stability and robust performance) [38],[39]. It must be pointed out that each block in the control diagram (Fig. 2.1b) can be represented by a *transfer function.* The transfer function relates the output of each block with respect to its input. For instance, the controller block relates the control command with respect to deviation error. This is, $C(s) = u(s)/\varepsilon(s)$.

2.2.1 The PI Controller

It is common to combine the proportional action with the integral action to get a classical PI controller. Let us discuss the design of a PI controller. Consider the first order nonlinear differential equation

$$\dot{x} = f(x) + d(t) + g(x)u \tag{2.5}$$

where $x \in \mathbb{R}$, $f: \mathbb{R} \to \mathbb{R}$ is a nonlinear functions, $d(t)$ is the external disturbing signal (which can be represented by the exciting force), $g: \mathbb{R} \to \mathbb{R}$ is the control input vector and u is the control command. Assume that one desires to lead the trajectories of the scalar $x(t)$ to the prescribed point $r = r^*$, i.e, the system output is $y = x$. In such a case, classical control theory proves that a control command $u = k_c(y-r) + (k_c/\tau_I)\int(y(\tau)-r(\tau))d\tau$ where k_c is the control gain and τ_I denotes the characteristic time of the feedback control, yields the asymptotic stabilization around the constant reference $r = r^* \in \mathbb{R}$ [39][40]. The above control command can be rewritten, from the Laplace operator as follows

$$u(s) = k_c\left(1 + \frac{1}{\tau_I s}\right)e(s) \tag{2.6}$$

where $e(s) = y(s) - r(s)$.

In this way the classical PI controller includes the following parts [33], [34]:

(i) *Proportional action.* The acting output of this part is proportional to the error $\varepsilon = y - r$, where y is the system output (measurement) and r is the reference (desired signal). Then, the proportional control action is given by: $u = u_s + k_c e$, where u_s is the bias signal of the controller (*i.e.*, the acting output of the controller when $e = 0.0$; namely, steady-state control command). A proportional control action is described by its proportional band (PB). PB is related with the control gain as follows: $PB = 100/k_c$. The PB characterizes the range over which the error must change in order to drive the acting signal of the controller over its full range. As smaller PB (or equivalently larger k_c) as larger sensitivity of control signal to deviations e.

(ii) *Integral action.* The control command includes a reset time (reset time is related to the integral time constant). The reset time is an adjustable parameter and sometimes it is also referred as *repeat time.* The integral action causes the control command to change as long as an error exists in the system output. Therefore, such control action can eliminate even small error if the reference is constant. In principle, locally speaking, the controller (2.6) stabilizes a dynamical system for any $k_{c,1} \leq k_c \leq k_{c,2}$ (where $k_{c,1}$ and $k_{c,2}$ are constant) if and only if the characteristic equation of the closed-loop system has all its roots in the open left-half complex plane (for more details see the control classical literature, [33],[34]).

If the desired signal is a constant value, the PI controller is able to eliminate small stationary errors (off set). In order to illustrate the capabilities of the PI controller in the chaos stabilization problem, consider the duffing system. The position has been chosen as system output (measured variable, $y = x_1$) and the closed-loop system becomes (readers can compare it with equation (2.2))

$$\begin{aligned} \dot{x}_1 &= x_2 + u \\ \dot{x}_2 &= x_1 - x_1^3 - \pi_1 x_2 + \alpha\cos(\omega t) \\ \dot{x}_3 &= y - r \end{aligned} \tag{2.7}$$

where $u \in \mathbb{R}$ is given by the time-domain equivalent representation of the equation (2.6), x_3 defines the integral error, $\pi_1 = 0.15$, $\alpha = 1.75$ and $\omega = 2/3$. In particular, the

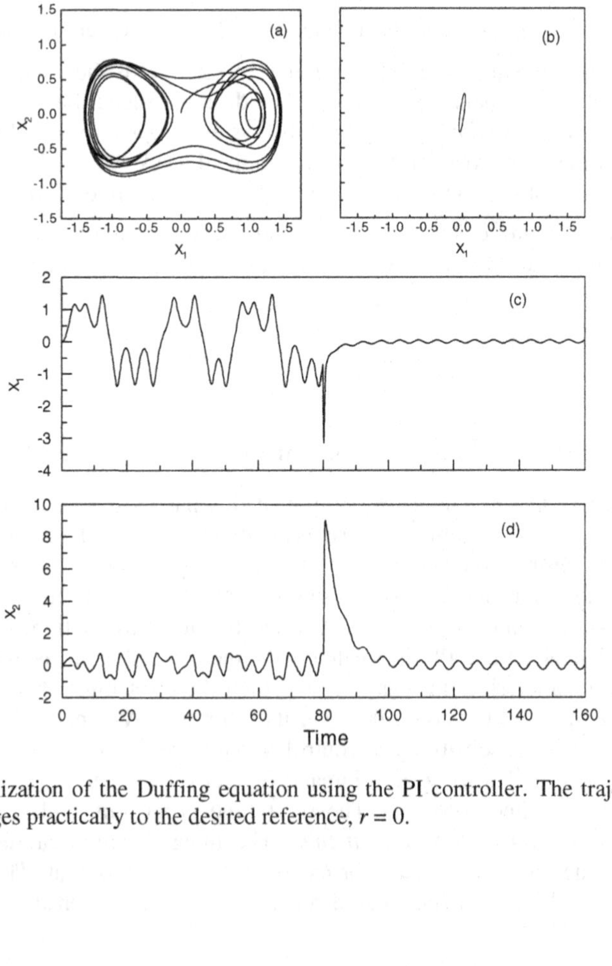

Fig. 2.7. Stabilization of the Duffing equation using the PI controller. The trajectories of the system converges practically to the desired reference, $r = 0$.

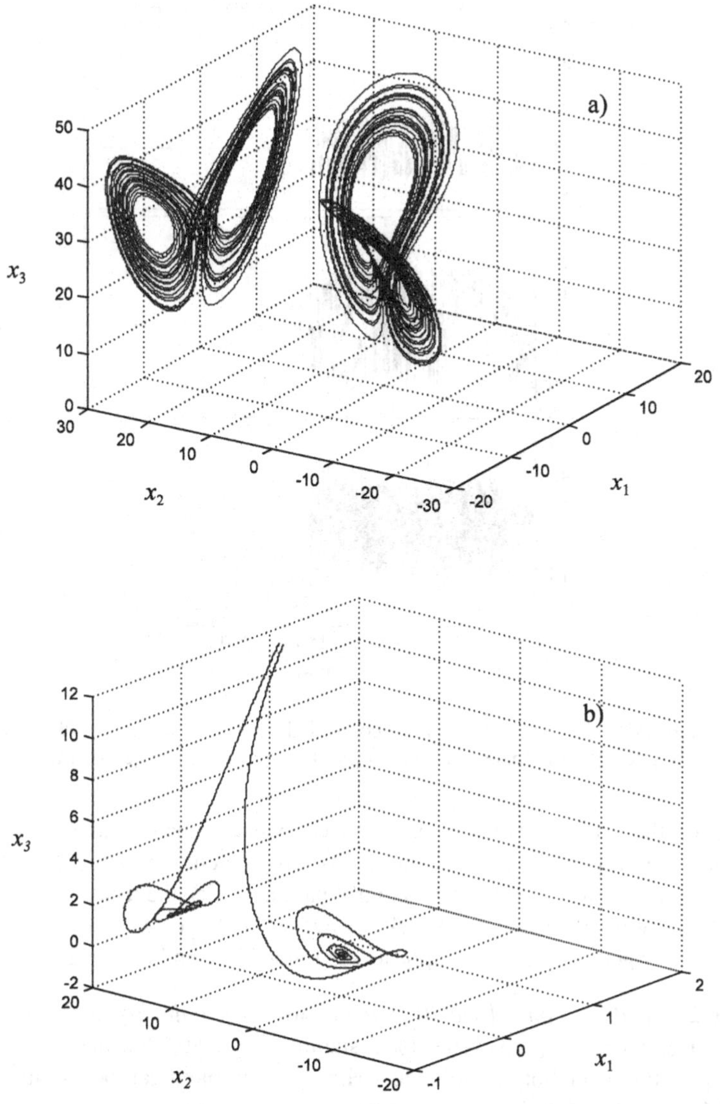

Fig. 2.8. Phase portraits of Lorenz equation, (a) control off and (b) control on

Duffing system can be stabilized at the origin for all $\infty > k_c \geq 0.86$ and any initial condition at the region $\Omega \subset \mathbb{R}^2$.

Note that, in spite of the control command only affects the first information channel, the velocity is also stabilized. In fact, as we shall see below, it is related to internal stability properties as, for example, minimum-phase. Figure 2.7 shows the stabilization of the Duffing equation at origin by means of the controller (2.6) for $k_c = 5.0$. The controller was activated at $t = 80$ ($u = 0$ for all $t \leq 80$ sec). Note that

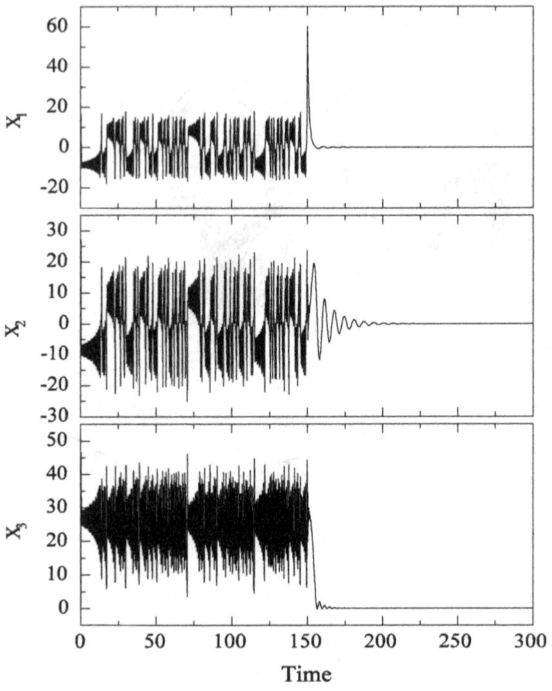

Fig. 2.9. High order chaotic system can be controlled via the PI feedback. PI control actions asymptotically steers the Lorenz equation to origin. $k_c = 150$ and the reset time $\tau_I = 1.0$.

structure of the Duffing attractor is removed, which means that chaos has been suppressed (See Figure 2.7a and Figure 2.7b). The system is practically stabilized at origin. Indeed, for larger value of the control parameter, k_c, the trajectory $x(t)$ tends to a ball whose radius is of magnitude $1/k_c$. However, as larger value of the control parameter k_c as larger overshoot can appear (see Figure 2.7c and Figure 2.7d).

Exercise 2.4. *Duffing control via PI feedback with low gain.* By choosing values of the control parameters k_c satisfying local stability $k_c \geq 0.86$, find the closed-loop behaviour such that overshot is minimized. Hint: The problem can be formulated as an Optimal Control Problem in the sense that we desire to find the value of k_c such that the supreme value of the control function (2.6) is minimized.

Chaos can be also suppressed even in high order systems by means of the PI controller [13]. Indeed, we can choose the first state as observable, $y = x_1$. In such way that the close-loop system is given by system (2.4) where u is given by the time-domain equivalent equation of the controller (2.6). The Lorenz system is stabilized at origin for all $k_c \geq 136$. Figure 2.8 and Figure 2.9 shows the stabilization at the origin for $k_c = 150$. The controller was activated at $t = 150$. Note that, as the control gain increases as the system frequencies collapse. The PI controller does not remove fundamental frequencies; it stabilizes orbits contained into the attractor.

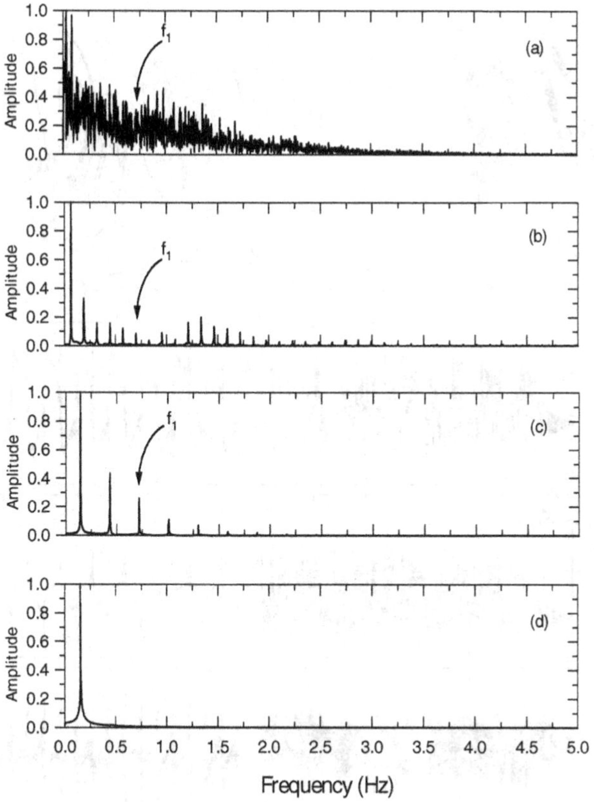

Fig. 2.10. Power spectrum of the Lorenz equation under PI control actions for several control gains. r = 0 and $\tau_I = 1.0$ were chosen. (a) $K_C = 0$, the controller is not activated, (b) $K_C = 10.0$, (c) $K_C = 50$, (d) $K_C = 136$ and (e) $K_C = 150$.

In seek of completeness and comparison with previous section, we compute the PSD, Figure 2.10 shows the power spectrum of the Lorenz equation, under control actions, for several control gain values. The spectra was obtained by means of FFT from the time series of the measured signal (3000 data were computed). Note that for all $0 < k_c \leq 50$ the chaos is not completely suppressed (see also Fig. 2.5). To compute time series, the controller was activated for all $t \geq 0$. Note that the controller is not a low pass filter. Indeed, signals collapse to derive a "new" signal. The "new" signal rises due to the stabilization of the unstable periodic orbits. For instance, the peak, whose frequency corresponds to 0.2 Hz, rises for $k_c \geq 50$ (Figure 2.10.c) while for $k_c \geq 10$ such frequency is not present into the spectra (see Figure 2.10b). Note also that the amplitude of the peak at $f_1 \approx 0.75$ Hz varies with control parameter k_c.

In some sense, the control gain may be interpreted as the cut frequency of a low pass filter [35]. However, a more appropriate interpretation is the following: *The feedback control (2.6) is an orbit stabilizer via the control gain.* This is, as the control gain increases a large number of unstable orbits can be stabilized. As a consequence

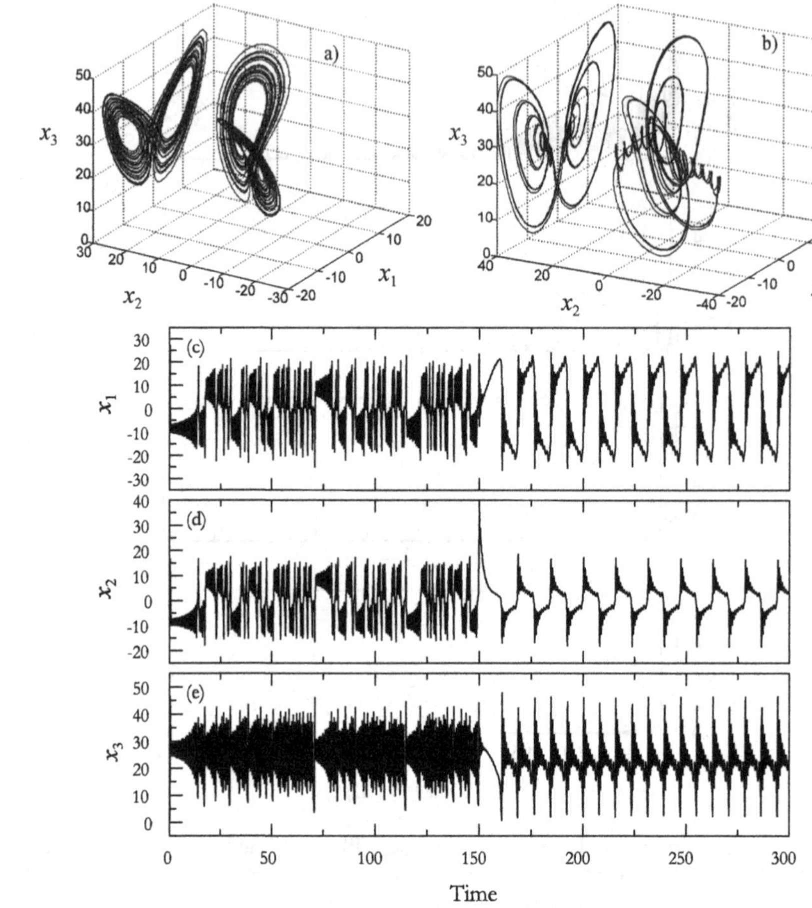

Fig. 2.11. Dynamic evolution of the Lorenz equation under PI control actions for k_c = 10. The stabilization of the UPO's by controller induces change of the attractor structure.

of the orbit stabilization, the closed-loop system has a smaller number of fundamental frequencies (see Figure 2.10).

For example, consider the closed-loop system (2.4). Now, let us consider that the control parameter k_c = 10. The closed-loop dynamics is showed in Figure 2.11. Note that the corresponding power spectrum (see Figure 2.10b) has less fundamental frequencies than k_c = 0 case. It is clear that, if k_c = 10.0, apparently the chaotic behaviour is not completely suppressed; however, some orbits have been stabilized. Thus, under control actions and k_c = 10.0, the attractor change its dynamical structure in such way that the projection onto the canonical plane (x_1, x_3) has an "*owl eyes*" structure. The change of the dynamical structure is due to the Proportional-Integral control actions.

The controller (2.6) yields the stabilization of a chaotic system at the origin. However, it is well established that a PI controller cannot solve the tracking problem [33],[34],[40],[41] for a reference signal with arbitrary frequency. The tracking problem consists on the design of a feedback control which can leads the system output

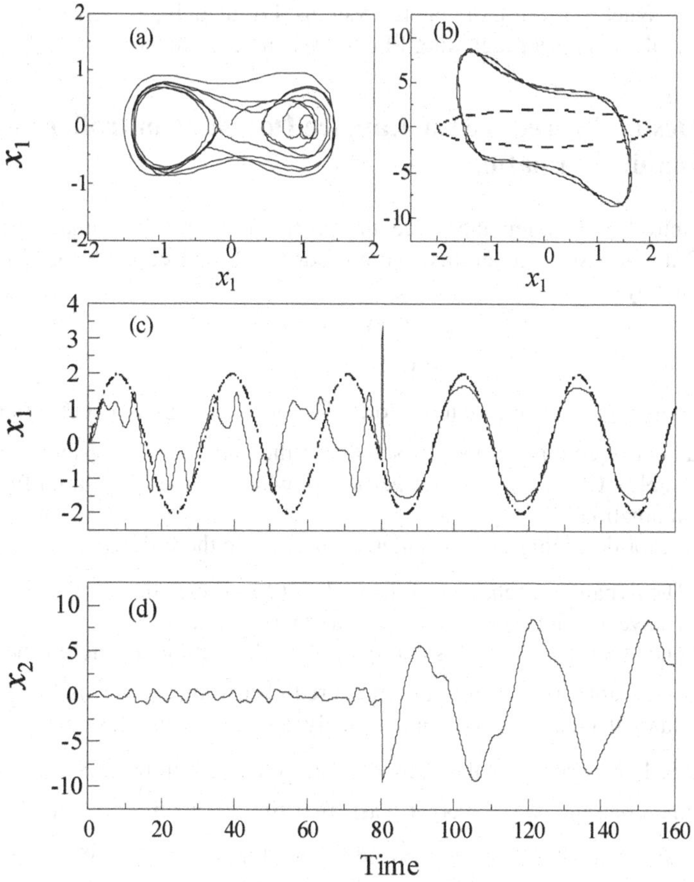

Fig. 2.12. Tracking of a time function (tracking problem) cannot be solved by a PI controller

into a time-varying desired signal. In other words, the reference signal is a time function. Thus, the tracking of the reference must be guaranteed by the designed feedback controller. Figure 2.12 shows the tracking of a periodic orbit by the Duffing equation using the same values of the control gain than Figure 2.7. It is clear that the reference cannot be attained by the controller (2.6) whereas stabilization at origin can be obtained. Hence, in order to stabilize a chaotic system around of a desired orbit, a modification of the classical PI controller is required.

In this manner, chaos suppression at points or (slow) orbits is possible with PI controllers. Moreover, if the synchronization problem can be stated as the stabilization of the discrepancy between drive and response systems, then the PI controller is able to yield the synchronization via the stabilization, see sec. 2 of Chapter 3. Nevertheless, as we can see in Figure 2.12, the classical PI cannot yield tracking of a time-function reference. Therefore, if the synchronization problem cannot be stated as a stabilization problem, then the classical PI controller does not yield chaos synchronization. In addition, if the relative degree [36] [37] is not equal to one, then the PI controller cannot yield synchronization of chaotic systems. Fortunately, as we shall see in Chapter 4, the results in present regarding PI control can be extended.

2.3 A Design Procedure for Laplace Domain Controllers with Dynamic Estimation

Here, we introduce a systematic procedure for the design of the Laplace domain controllers. Without lost of generality, we consider a second order dynamical (chaotic) system given by

$$\begin{aligned} \dot{x} &= f(x) + g(x)u \\ y &= Cx \end{aligned} \tag{2.8}$$

where $x \in \mathbb{R}^2$ represents the state vector, $f : \mathbb{R}^2 \to \mathbb{R}^2$ is a smooth vector field, $u \in \mathbb{R}$ denotes a control command, $y \in \mathbb{R}$ stands for the output of the system (i.e., a measured observable), C is a vector of proper length which defines the output function y as a linear combination.

Now, for designing the controller, let us assume the following

S.2.1) The dynamical behavior of the system (2.8) is chaotic.
S.2.2) The vector field $f(x)$ is unknown and unobservable.
S.2.3) The control command is acting on the information channel of the measured observable. In addition, the measured observable, $y \in \mathbb{R}$ is function of one state, that is, only one state of the dynamical system (2.8) is measurable.

Lemma 2.5. Let us consider the second order system given by (2.8). Suppose that $y \in \mathbb{R}$ is time continuous and has, at least, first time derivative. If a nonlinear affine system x $= f(x) + g(x)u$ (where $x \in \mathbb{R}^2$, $f(\cdot)$ and $g(\cdot)$ are continuous vector fields and $u \in \mathbb{R}$) has relative degree well posed and equal to 1, then it can be written in the following form

$$\begin{aligned} \dot{z} &= \phi(z) + \gamma(z)u \\ \dot{w} &= \Omega(z, w) \end{aligned} \tag{2.9}$$

where z, u $w \in \mathbb{R}$, φ(z), γ(z) and Ω(z,w) are continuous (possibly nonlinear) functions. ■

Remark 2.6. Assumption S.2.1) implies the trajectories of (2.8) asymptotically converge to an attractor for any initial conditions $(z(0),w(0)) \in \mathbb{R}\times\mathbb{R} \subseteq \mathbb{R}^2$. That is, the state $(z,w) \in \mathbb{R}$ is globally bounded. In addition, note that vector field $f(z,w)$ includes "internal" and "external" (interactions) forces; given by w and z, respectively. Since, by assumption, the trajectories of the system (2.8) asymptotically converge to a chaotic attractor. Hence, also the trajectories (z(t),w(t)) asymptotically converge to a chaotic attractor. Assumption S.2.2) involve that the dynamical model of the system (2.8) is not perfect. Assumption S.2.3) implies that the relative degree of system (2.8) is equal to one [36] and that the output function is a linear combination of the state variables. Note that the proof of Lemma 2.1 is straightforward.

Remark 2.7. In this section we consider the input vector field $g(x) = [1\ 0]^T$. This implies that $\gamma(z) = 1.0$ which means the vector field related with control command is known. This assumption can be satisfied for any chaotic circuits. However, such a situation cannot be, in general, satisfied. Therefore such an assumption shall be relaxed in Sect. 2.4. Indeed, Assumptions S.2.1-S.2.3 shall be relaxed along the text toward generalization. ○

Lemma 2.8. Let us suppose that $\gamma(z) = 1.0$. If, for a positive constant β and a class $\mathcal{K}$ function α, $\| \Omega(z,w) \| \leq \alpha(\| z_\tau \|)_{\mathcal{L}} + \beta$, then the feedback $u = -[\phi(z) + k(z - z^*)]$ leads the trajectories of (2.9), $(z(t),w(t))$, to a fixed point $(z^*,w^*) \in \mathbb{R}\times\mathbb{R}$ for any initial condition $(z(0),w(0)) \in \mathbb{R}\times\mathbb{R} \subseteq \mathbb{R}^2$.

Proof. Defining the stabilization error $e = (e_1,e_2)$ where $e_1 = z - z^*$ and $e_2 = w - w^*$ and (z^*,w^*) are the coordinates of the prescribed reference. This implies that $e^* = (0,0)$ is an equilibrium point (i.e., $z = z^*$ and $w = w^*$). Under control actions (closed-loop), the dynamics of the stabilization error is as follows

$$\begin{aligned} \dot{e}_1 &= -ke_1 \\ \dot{e}_2 &= \Omega(e_1,e_2) \end{aligned} \tag{2.10}$$

Since, the trajectories of system (2.9) asymptotically converge to an attractor, hence $w(t)$ is bounded and so $\Omega(e)$. In consequence, for any positive defined constant k, sufficiently large, the trajectory of $e_1(t)$ converges to zero, which implies that $z \rightarrow z^*$ for all $t > 0$ and any $z(0)$ in the region R. By assuming that $\| \Omega(z,w) \| \leq \alpha(\| z_\tau \|)_{\mathcal{L}} + \beta$, for any positive constant β and a class $\mathcal{K}$ function α, we have that the subsystem $e = \Omega(z,w)$ is $\mathcal{L}$-stable. Therefore, system (2.9) is minimum-phase. Hence, see [36], the system (2.9) under the control $u = -[\phi(z) + k(z - z^*)]$ is asymptotically stable at the equilibrium point (z^*,w^*). ■

Remark 2.9. According to assumptions *S.2*) and *S.3*), the vector field is unknown. Therefore the continuous function, $\phi(z)$ is also unknown. Hence the feedback controller $u = -\phi(z) + k(z - z^*)$ cannot be directly implemented because it needs a priori knowledge about $\phi(z)$. Then, a modification of such controller must be developed in order to account lack of knowledge on the vector field $f(x)$. ○

Theorem 2.10. Let us define $\eta = \phi(z))$ and $\psi(z,w,\eta) = \eta - \phi(z)$. If $\psi(z(0),w(0),\eta(0)) = 0$ at $t = 0$, then system (2.9) can be rewritten in the following dynamical equivalent form

$$\begin{aligned} \dot{z} &= \eta - gu \\ \dot{\eta} &= \Gamma(z,w,\eta) \\ \dot{w} &= \Omega(z,w) \end{aligned} \tag{2.11}$$

Proof. Suppose that $\psi(z(0),w(0),\eta(0)) = 0$. This implies that the manifold $\psi = \{(z,w,\eta) \in \mathbb{R}^3 : \eta - f(z) = 0\}$ is invariant under the trajectories of (2.11). This means that $\psi(z,w,\eta)$ is a first integral of the system (2.11). Hence, the system (2.11) is dynamically equivalent to the system (2.9). ■

Remark 2.11. The systems (2.11) and (2.9) are dynamically equivalent. Hence a designed feedback for (2.11) can lead the trajectories of the system (2.9) to the prescribed reference (z^*,w^*) in face of lack of knowledge if the initial condition satisfies $\psi(z(0),w(0),\eta(0)) = 0$. In fact, the feedback control $u = -[\eta + k(z - z^*)]/\gamma = -[\phi(z) + k(z - z^*)]/\gamma$ due to by definition, $\eta = \phi(z)$. However, since the system (2.9) has a relative degree, $\phi = 1$, the *new* state η is not available for feedback. Indeed, it cannot be measured because it is an intermediate variable toward the final controller which defines the dynamical behavior of the uncertain vector field $f(x)$. ○

Theorem 2.12. If the systems (2.9) and (2.11) are dynamically equivalent, then the controller leads the trajectories of system (2.9) asymptotically to the prescribed reference (z^*,w^*).

$$
\begin{aligned}
\dot{\hat{z}} &= \hat{\eta} + \gamma(z)u + L\kappa_1(z - \hat{z}) \\
\dot{\hat{\eta}} &= L^2\kappa_2(z - \hat{z}) \\
u &= \frac{1}{\gamma(z)}\left[\hat{\eta} + k(\hat{z} - z^*)\right]
\end{aligned}
\tag{2.12}
$$

Proof. Following the procedure reported in [2], the above result is straightforward (also see appendix A in [42]). ■

Remark 2.13. If the term $\gamma(z)$ does not depend on the state variable or is constant (such conditions are often satisfied in chaotic circuits), then the controller (2.12) is linear. Thus Laplace transform is straightforward. Hence the transfer function $C(s) = u(s)/e(s)$, can be obtained

$$
u(s) = k_c\left[1 + \frac{1}{\tau_I s} + \frac{k_e}{s(s+\kappa_1)}\right]e(s) \tag{2.13}
$$

from where we have that, by adding high-order integral actions, an uncertain nonlinear system in form (2.9) can be stabilized at the point (z^*,w^*) despite the unknown vector field $f(x)$. ○

Corollary 2.14. The feedback controller (2.13) asymptotically steers the trajectories of the system (2.9) to the prescribed reference (z^*,w^*). □

Note that the controller (2.13) does not require information a priori about the system to be controlled. In fact, it only requires: (i) Knowledge about the reference signal and (ii) On-line measurements of any available state. In consequence, the control law (2.13) can be directly implemented to control any system with bounded perturbing forces. It must be pointed out that the controller (2.13) contains three parts: (i) The proportional action, (ii) the integral action and (iii) a quadratic integral action. Due to this structure, the controller (2.13) was called PII^2. The two first actions (Proportional-Integral) were previously described. The third one (Quadratic integral action) provides a dynamic estimation of the internal perturbing forces; in such way that the system attains the reference signal even in high frequencies (this is, $t \to T$, where T is the time when the control is activated). Let us define the following control error: $e_1 = y - \bar{y}$ and $e_2 = d - \bar{d}$. Then, dynamics of the control error is governed by: $\dot{e} = A(\kappa)e + \Phi(t)$, where $\Phi(t) = [0, F(t)]^T$ and the companion matrix $A(\kappa)$ is given by

$$A(\kappa) = \begin{bmatrix} -\kappa_1 & 1 \\ -\kappa_2 & 0 \end{bmatrix}$$

Hence, the control error, $e = (e_1,e_2)$, converges asymptotically to the point $e = (0,0)$ for all $\kappa_1, \kappa_2 > 0$ if and only if the internal perturbing force and the dynamics is smooth and bounded [43]. Fortunately, the most chaotic system satisfies this condition. In consequence, chaos control can be physically implemented. The controller (2.13) has been obtained from model. Nevertheless, the Laplace control can be designed in a pure form. That is, by taking an measured state and comparing its actual value with the desired value the control error is computed and, by using eq. (2.13), the control command is calculated. Thus, in order to extend results in inductive manner, let us now consider that the nonlinear system to be controlled is given by the Lorenz equation. Assume that the measured state is $y = x_1$ which is the same control information channel. Then, the closed-loop system is given by (2.4) where u is given by the time domain equivalent equation of the controller. This is, using the inverse Laplace transform, the control law (2.13) becomes: $u = u_S + k_c\,[1 + k_e \exp(-t/\tau_{II})](y - r) + k_c\,\tau_I^{-1}\eta$. The integral variable, η, provides the dynamics of the regulation error.

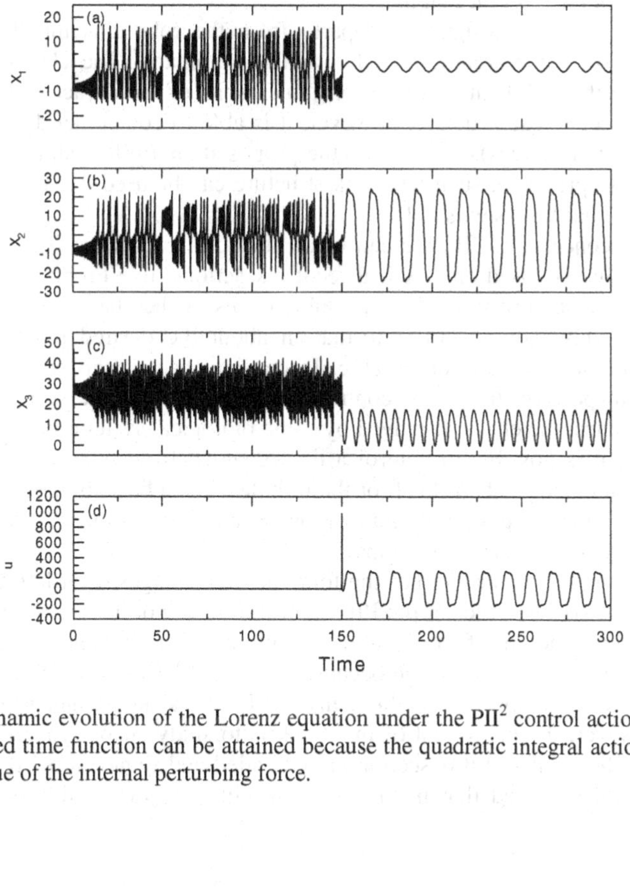

Fig. 2.13. Dynamic evolution of the Lorenz equation under the PII^2 control actions. The tracking of a desired time function can be attained because the quadratic integral action provides an estimated value of the internal perturbing force.

The reference signal was selected as the solution of the following oscillator

$$\begin{aligned} \dot{r}_1 &= r_2 \\ \dot{r}_2 &= -\omega_r r_1 \end{aligned} \tag{2.14}$$

where $\omega_r > 0$ denotes the natural frequency of system (2.14). Then since solution of system (2.14) is periodic the reference signal becomes $r(\mathrm{t}) = \alpha\cos(\omega_r \mathrm{t})$, where the amplitude α is directly related to initial conditions $r(\mathrm{t} = 0) \in \mathbb{R}^2$. Figure 2.13 shows the tracking of the reference signal. The control parameter were chosen as follows: $\mathrm{k_c} = 150$, $\mathrm{k_e} = 100$, $\tau_{\mathrm{I}} = 1.0$, $\tau_{\mathrm{II}} = 1.0$. The control was activated at $\mathrm{t} = 150$ (which means that $u = 0$ for all $\mathrm{t} \leq 150$). Note that the dynamics of the control command, u, is not chaotic even the chaotic behavior of the Lorenz system has been suppressed. Moreover, in spite of the control command only affects x_1; x_2, and x_3 display regular behavior. Note that the Lorenz equation (2.4) is in some sense synchronized with the periodic oscillator (2.14) via the PII^2 controller. Such synchronization is achieved by *tracking* the reference $r(\mathrm{t})$. That is, the time function $r(\mathrm{t})$ is yielded by a reference system (which can be so-called drive system); then under closed-loop, the output of the response system tracks the reference signal. The output of the response system (2.4) is synchronized with the output of the driven system (2.14). In some manner if the driven system can be chosen as a chaotic system, the output of the response system should track the reference signal.

The Laplace domain design can depart from the robust asymptotic controller and/or the classical interpretation of the control command. The classical feedback actions provide the stabilization of the unstable orbits of the attractor. The resulting controller has a classical structure; however, it is able to counteract dynamic signals (bounded perturbing forces). In this way the proposed controller can tracks a desired time function. In principle, such feedback structure can be used to carry out the synchronization of chaotic systems [44].

In addition, concerning PII^2, we can claim the following facts: (i) The controller (since its structure is based on classical feedback actions; like Proportional and Integral) provides the stabilization of the unstable orbits of the chaotic attractor and (ii) The controller only requires: (a) information about the desired reference and (b) on-line measurements of any observable. In consequence, it can be experimentally implemented to perform the chaos control on high-order chaotic system from time series. For instance, the chaos displayed by some biological systems can be controlled [45],[46] (e.g., it is possible to control a thorax signal from time series, which has been shown that displays chaos [47]) or the turbulence can be stabilized (for instance, chaos suppression in a gas-solid fluidization system [48], which is, under certain conditions, a high-order chaotic systems).

In addition, the PII^2 controller has the following advantages regarding the adaptive control schemes: (i) The order of the PII^2 controller does not increases with the number of parameters and (ii) If the system is nonlinear in its parameter structure, the proposed controller does not change because the controller does not require information about system parameters. In this sense, PII^2 offers an advantage on adaptive schemes. However, the main goal of this book is to study chaos synchronization. In what follows, the results in this section are extended and foundational theory is presented. It should be noted that in this book synchronization is addressed from the stabilization of the discrepancy between drive and response systems. Indeed, synchronization via PI controller is detailed until Chapter 3.

2.4 Rationale of Chaos Stabilization Via Robust Geometrical Control

In previous section we show how the model-based design can lead us to Laplace domain controllers. Some heuristic notion were used to extend results from 2-dimensional system to 3-dimensional ones. Now, the rationale of the section is extended to design feedback controller for chaos control. Formalization for n-dimensional systems is dealt in Chapter 4; here the chaos suppression in 2 and 3-dimensional systems is discussed.

Essential elements in chaos suppression are the following: (a) Control of 'erratic' dynamics in a given system and (b) stabilization of a chaotic system about a given reference trajectory. In this section, we present an input-output controller based on geometrical control theory [36],[37] for low dimensional (2D and 3D) systems. The main idea is to lump the uncertainties in a nonlinear function which can be interpreted as a new state in an externally dynamically equivalent system. Thus, the new state is estimated by means of a high-gain state observer. The state observer provides the estimated value of the lumping nonlinear function (and consequently of the uncertainties) to the linearizing feedback control. As we shall see below, this design algorithm can yield, under certain conditions, in feedback control as in previous section.

Consider the following class of nonlinear system whose trajectories are contained into a chaotic attractor

$$\dot{x} = f(x) + g(x)u \tag{2.15}$$

(where $x \in \mathbb{R}^n$, for n =2 or 3, is a state vector, $u \in \mathbb{R}$ is a scalar input, $f(x)$ and $g(x)$ are smooth vector fields). Assume that $y = h(x) \in \mathbb{R}$ is the system output ($h(x)$ is a smooth function) and consider $L_f h(x) = \partial h/\partial x f(x) = [\partial h/\partial x_1\ \partial h/\partial x_2 \ldots \partial h/\partial x_n][f_1\, f_2 \ldots f_n]^T$. If ρ is the smallest integer such that the following conditions hold at $x = x_0$, the above system is said to have a relative degree ρ at x_0. The conditions are: (i) $L_g L_f^i h(x) = 0$, $i = 1,2, , \rho -2$ and (ii) $L_g L_f^{\rho-1} h(x) \neq 0$, (where $L_f^i h(x) = L(L_f^{i-1} h(x))$, $L_f h(x)$ is the Lie derivative of $h(x)$ along $f(x)$) [36]. If the smoothness assumption and the above conditions are satisfied, then by defining ρ new coordinates (which is a diffeomorphism), $z_{i+1} = L_f^i h(x)$, for $i = 0,1, , \rho -1$, the above system can be written in the following canonical form [36],[37]

$$\begin{aligned}
\dot{z}_i &= z_{i+1} \qquad i = 1,2,\ldots,\rho - 1 \\
\dot{z}_\rho &= \alpha(z,v) + \gamma(z,v)u \\
\dot{v} &= \varsigma(z,v) \\
y &= z_1
\end{aligned} \tag{2.16}$$

where $\alpha(z,v) = L_f^\rho h(x)$ and $\gamma(z,v) = L_g L_f^{\rho-1} h(x)$. Thus, the following feedback $u = -[L_f^\rho h(x) + V(x)]/L_g L_f^{\rho-1} h(x)$ is a linearizing control law. In fact, according to results in Chapter 4 of [36], the global asymptotic stabilization of system (2.16) can be achieved by means of the above feedback. In addition, same results can be extended for tracking problem. In this way, the coordinates exchange $z = \Phi(x) := L_f^i h(x)$, allows to design a feedback control for a nonlinear system (2.15) given by

$$u = -[\, L_f^{\rho} h(x) + V(x)]/L_g L_f^{\rho-1} h(x) \tag{2.17}$$

The feedback (2.17) has the following features: it is able to induce linear behavior into system (2.16) and as a consequence, into the system (2.15). The controller (2.17) counteracts the nonlinear terms $\alpha(z,v)$ and $\gamma(z,v)$ and induces the linear behavior to stabilize at origin $V(x) = K^T z$. The controller (2.17) can be computed for finite number of measured states $y = h(x)$. That is, it is not required to measure all states of system (2.16). Nevertheless, feedback (2.17) requires exact knowledge about the vector fields $f(x)$ and $g(x)$ and the Lie derivatives of the system output $h(x)$. This situation is not realistic for controlling chaos. In such a problem the vector fields are uncertain and the system output is unpredictable. For example let us consider the problem of synchronization between chaotic systems. If we think that: (i) chaotic systems posses sensitive dependence on initial conditions and it is almost impossible to reproduce the same starting condition on two or more chaotic systems, (ii) even infinitesimal variations of any parameter value will eventually results in divergence of chaotic systems starting nearby each other, and (iii) it is expected that mathematical model of the drive system is unknown, then it is not hard to see that the above feedback requires much information to be physically implemented for synchronizing chaos. Indeed, such feedback should be modified in order to be used to control chaos. Such modification makes sense only if main features of controller (2.17) remains into new feedback.

Without loss of generality, we can suppose that the reference (constant) signal is $r = 0$. Then, in coordinates (z,v) the linearizing feedback control becomes $u = [\alpha(z,v) + K^T z]/\,\gamma(z,v)$, where K's are such that the polynomial $s^{\rho} + k_{\rho} s^{\rho-1} + \ldots + k_2 s + k_1$ is Hurwitz. Nevertheless, if the vector fields $f(x)$ and $g(x)$ are uncertain, the coordinates transformation $z = T(x)$ bringing the original system into the canonical form (2.16) is uncertain even in Lie derivatives exists. In principle, since the coordinates transformation is a diffeomorphism, one can suppose that: i) The uncertain transformation exists and, ii) it is invertible. That is, since the nonlinear functions $\alpha(z,v)$ and $\gamma(z,v)$ are uncertain, $T(x)$ is also uncertain. Moreover, the linearizing control law has been designed in such a manner that all the observable states $z \in \mathbb{R}^{\rho}$ are needed for feedback. We use the linearizing feedback (2.17) only as an intermediate control law toward the modified controller. Towards the robust approach; to this end, let us assume the following

S.2.4) Only the system output $y = z_1 \in \mathbb{R}$ is available for feedback.
S.2.5) $\gamma(z,v)$ is bounded.
S.2.6) The system (2.16) is minimum phase.
S.2.7) The nonlinear functions $\alpha(z,v)$ and $\gamma(z,v)$ are uncertain. However, an estimate $\hat{\gamma}(z)$ of $\gamma(z,v)$, satisfying $\mathrm{sign}(\hat{\gamma}(z)) = \mathrm{sign}(\gamma(z,v))$, is available for feedback.

Now, let us define $\delta(z,v) = \gamma(z,v) - \hat{\gamma}(z)$, $\Theta(z,v,u) = \alpha(z,v) + \delta(z,v)u$ and $\eta = \Theta(z,v,u)$. In addition, let us consider the following dynamical system

$$
\begin{aligned}
\dot{z}_i &= z_{i+1} \qquad 1 \le i \le \rho - 1 \\
\dot{z}_\rho &= \eta + \hat{\gamma}(z)u \\
\dot{\eta} &= \Xi(z,\nu,\eta,u) \\
y &= z_1
\end{aligned}
\tag{2.18}
$$

where $\Xi(z,\eta,\nu,u) = \sum_{k=1}^{\rho-1} z_{k+1}\,\partial_k\Theta(z,\nu,u) + [\eta + \hat{\gamma}(z)u]\partial_\rho\Theta(z,\nu,u) + \delta(z,\nu)u + \partial_\nu\Theta(z,\nu,u)\zeta(z,\nu)$, $\partial_k\Theta(z,\nu,u) = \partial\Theta(z,\nu,u)/\partial x_k$.

Proposition 2.15. The manifold $\Psi(z,\eta,\nu,u) = \eta - \Theta(z,\nu,u) = 0$ is invariant under the trajectories of the system (2.18).

Proof. It suffices to prove that $d\Psi(z,\eta,\nu,u)/dt = 0$ along the trajectories of system (2.18) which, using the definition $\eta = \Theta(z,\nu,u)$, is straightforward. □

Proposition 2.16. The system (2.18) is dynamically externally equivalent to the system (2.16). This is, for all differentiable input $u \in \mathbb{R}$. The system (2.18) has the same solution as the system (2.16) module $\pi\cdot(z,\eta,\nu) \to (z,\nu)$.

Proof. From the equality $\Psi(z,\eta,\nu,u) = 0$ and the condition $d\Psi(z,\eta,\nu,u)/dt = 0$, one can take the first integral [49] of system (2.19) to get $\eta = \Theta(z,\nu,u)$. When the first integral is back-substituted in system (2.18), we obtain the solution of the system (2.16). This implies that the solution $z(t) \in \mathbb{R}^\rho$ of the system (2.16) is the solution of the upper subsystem (2.18), hence $\pi(z,\eta,\nu) = (z,\nu)$. □

Exercise 2.17. Let us consider *the Hodgkin-Huxley neuron* (HH-neuron). The dynamical model of the HH-neuron is given by the following ODÉs system (for parameters values see [50] and references therein): $\dot{x}_1 = -\pi_1 x_1^4(x_1 - V_k) - \pi_2 x_3^3 x_4(x_1 - V_{Na}) - \pi_3(x_1 - V_l) + u$; $\dot{x}_2 = \alpha_2(1 - x_2) - \beta_2 x_2$; $\dot{x}_3 = \alpha_3(1 - x_3) - \beta_3 x_3$; $\dot{x}_4 = \alpha_4(1 - x_4) - \beta_4 x_4$ (where x_1 stands for the membrane potential, x_2, $x_3$and x_4 are related to gating variables for the sodium and potassium channels). The parameters πś are constant. By taking x_1 as system output, the system (2.16) can be straightly obtained. Show that by defining the lumping function $\Theta(z,\nu,u)$ the HH-neuron model can be extended to the form (2.18) and the both model and extended for have same simulation. Hint: Use numerical simulations by taking parameters and initial conditions for HH-neuron model as in [50] and chose initial conditions for the extended system with different value than model but satisfying Proposition 2.15. Thus, compare the $\Theta(z(t),\nu(t),u)$ with $\eta(t)$. ○

Remark 2.18. We can note the following from Exercise 2.13 : (i) The augmented state, η, provides the dynamics of the uncertain function $\Theta(z,\nu,u)$, and consequently of the uncertain terms $\alpha(z,\nu)$ and $\gamma(z,\nu)$. From the minimum-phase assumption the following result is not hard to prove. (ii) The relative degree of the HH-neuron model under supposition x_1 is the system output. Thus, HH-neuron illustrates that control in fourth order system is possible for the case we take x_1 as system output in similar manner 2 and 3-dimensional systems because of HH-neuron system has relative degree equal to 1and it is minimum phase. The discussion about high-dimensional system with relative greater than 1 is in next chapters.

Proposition 2.19. Under the feedback $u = (-\eta + K^T z)/\hat{\gamma}(z)$, where K's are the coefficients of a Hurwitz polynomial, the states of the system (2.18) converge asymptotically to zero.

The Proposition 2.19 will be demonstrated in Chapter 4. An important advantage of the system (2.18) is the following: The dynamics of the states (z,η) can be reconstructed from the output [51]. We can propose the following observer

$$\begin{aligned} \dot{\hat{z}}_i &= \hat{z}_{i+1} + L^i \kappa_i (z_1 - \hat{z}_1) \qquad 1 \le i \le \rho - 1 \\ \dot{\hat{z}}_\rho &= \hat{\eta} + \hat{\gamma}(\hat{z})u + L^\rho \kappa_\rho (z_1 - \hat{z}_1) \\ \dot{\hat{\eta}} &= L^{\rho+1} \kappa_{\rho+1} (z_1 - \hat{z}_1) \end{aligned} \tag{2.19}$$

where $(\hat{z},\hat{\eta})$ are estimated value of (z,η), respectively. Note that the uncertain term, $\Xi(z,\eta,v,u)$, has been neglected in the construction of the observer (2.19).

Theorem 2.20. Let $e \in \mathbb{R}^{\rho+1}$ be an estimation error vector whose components are defined as follows: $e_i = L^{\rho-i}(z_i - \hat{z}_i)$, $i = 0,1,\ldots,\rho-1$, and $e_{\rho+1} = \eta - \hat{\eta}$. For a sufficiently large value of the high-gain parameter L, the dynamics of the estimation error, e, converges asymptotically to zero.

Proof. Combining systems (2.19) and (2.18), the dynamics of the estimation error can be written as follows: $\dot{e} = LA(\kappa)e + \Gamma(z,\eta,v,u)$, where $\Gamma(z,\eta,v,u) = [0, \Xi(z,\eta,v,u)]^T$ and the companion matrix is given by

$$A = \begin{bmatrix} -\kappa_1 & 1 & 0 & \cdots & 0 \\ -\kappa_2 & 0 & 1 & \cdots & 0 \\ \vdots & \vdots & \vdots & \ddots & \vdots \\ -\kappa_\rho & 0 & 0 & \cdots & 1 \\ -r\kappa_{\rho+1} & 0 & 0 & \cdots & 0 \end{bmatrix} \tag{2.20}$$

where $r = \gamma(z,v)/\hat{\gamma}(z)$. The matrix (2.20) is Hurwitz if $r > 0$ for all $t \ge 0$ and κ's are positive defined. According to Assumption (A4), this condition is satisfied. In addition, since the trajectories $x(t)$ are contained in a chaotic attractor, hence $\Gamma(z,\eta,v,u)$ is bounded. Consequently, for any $L > L^* > 0$, $e(t) \to 0$ as $t \to \infty$ which implies that $(z,) \to (z,\eta)$. □

Corollary 2.21. Now, consider the following linearizing-like control law: $u = [-\hat{\eta} + K^T\hat{z}]/\hat{\gamma}(z)$. Under the above feedback the system (2.19) is asymptotically stable for $L > L^* > 0$.

Remark 2.22. High-gain observers can induce undesirable dynamics effects such as the peaking phenomenon [52]. To diminish these effects, the control law can be modified by means of

$$u = Sat\{[-\hat{\eta} + K^T\hat{z}]/\hat{\gamma}(z)\}$$

where $Sat: \mathbb{R} \to \mathcal{B}$ is a saturation function and $\mathcal{B} \subset \mathbb{R}$ is a bounded set [52] and [56]. Thus as the saturation of the controller is feedback, the antireset windup is recovered (see below for an example).

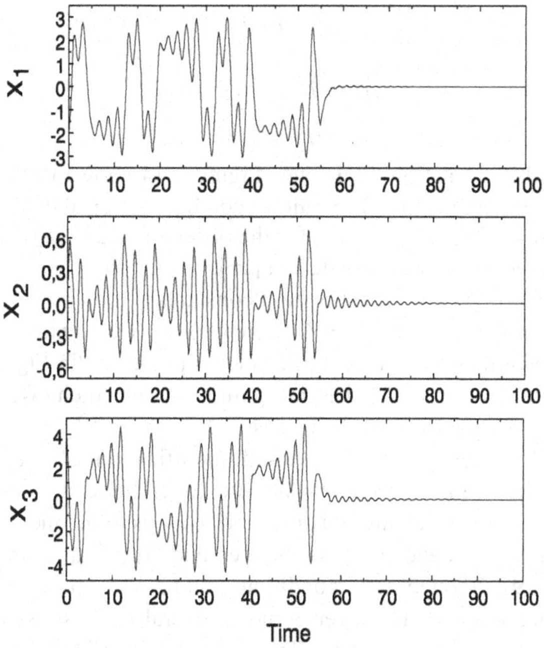

Fig. 2.14. Chaos suppression for the Chua's oscillator: L = 30

Let us consider the Chua's Circuit, which is a benchmark system for chaos suppression and synchronization [53], it is simple for realization even in inductorless configuration [54]. Synchronization is discussed on next chapter, here we illustrate the design of the robust feedback for suppressing chaos. The circuit equations can be written in dimensionless form as follows [25]

$$\begin{aligned}\dot{x}_1 &= \pi_1\left[x_2 - x_2 - \varphi(x_1)\right] + u\\ \dot{x}_2 &= x_1 - x_2 + x_3\\ \dot{x}_3 &= -\pi_2 x_2\end{aligned} \tag{2.21}$$

where $f(x_1) = bx_1 + ½(a - b)[\,|x_1 + 1| - |x_1 - 1|\,]$. In addition let us assume that the system output is given by $y = x_1$. Defining the invertible change of coordinates: $z_1 = x_1$, $v_1 = x_2$ and $v_2 = x_3$, the dynamical system (2.21) can be transformed into the canonical form (2.16) and its equivalent form (2.19) (with $\eta = \Theta(z_1,v)$ as the augmented state). Note that z_1 is the voltage across the capacitor C_1 which is bounded. Thus, the zero dynamics can be written as $= Cv + Dz_1$, where $D = [1,0]^T$ and

$$C = \begin{bmatrix} -1 & 1 \\ -\pi_2 & 0 \end{bmatrix} \tag{2.22}$$

which is Hurwitz if $\pi_2 > 0$ since π_2 is related to inductance, such condition is satisfied. Hence the system (2.21) is minimum-phase. Note that the Assumptions *S.2.4-S.2.7* are satisfied. Thus, the asymptotic controller becomes

$$\dot{\hat{z}}_1 = \hat{\eta} + u + L\kappa_1(z_1 - \hat{z}_1) \tag{2.23}$$

$$\dot{\hat{\eta}} = +L^2\kappa_2(z_1 - \hat{z}_1)$$
$$u = -\hat{\eta} + k_1\hat{z}_1 \tag{2.24}$$

where L > 0, is the estimation parameter. Figure 2.14 shows the stabilization of the Chua oscillator at the origin. The initial conditions for the system (2.21) were $(x_1(0),x_2(0),x_3(0))$ = (-2.0,0.02,4.0) and for the observer (2.23) $(z_1(0),(0))$ = (1,15). The model parameters values were chosen as in [25].

The control gain k_1 = 1.0, the estimation constants (κ_1,κ_2) = (2.0,1.0) and the high-gain estimation parameters value L = 30 were chosen. The controller was activated at t = 55.0s. The performance of the control input is presented in Figure 2.15. Note that the overshoot can be induced by the high-gain observer. Such effect increases as the value of the high-gain parameter L increases.

It is not hard to see that the proposed modification of the feedback control (2.23),(2.24) results in a PII^2 [13] controller due to (i) the relative degree of the open-loop system is one, i.e., the control channel corresponds to the measured state, and (ii) the uncertain terms $\alpha(z,v)$ and $\gamma(z,v)$ are lumped and after that, estimated via the high-gain observer (2.19) although, such conditions can be restrictive, they can be satisfied in many practical situations. However, a more generalized results about the proposed modification is presented in Chapter 4. The main idea is to lump the uncertainties in a nonlinear function which can be interpreted as an augmented state in a dynamically equivalent nonlinear system. A state estimator provides an estimated value of the augmented state, and consequently of the uncertainties.

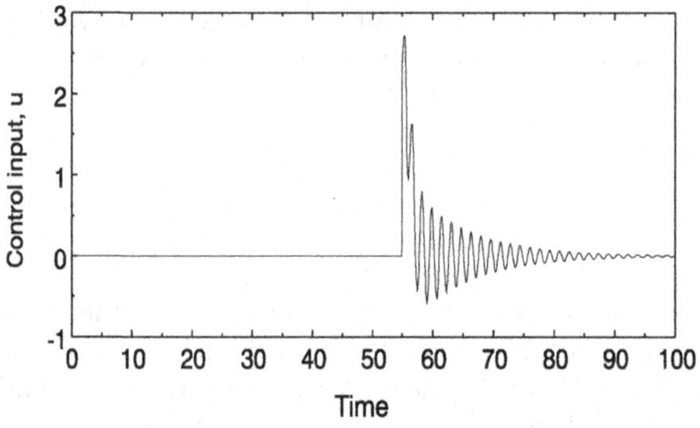

Fig. 2.15. Performance of the control, input without saturation

Now, all practical control systems can be subject to constraints on the manipulated inputs which are imposed due to the limited capacity of the actuators (valves, pumps, etc.). If such constraints are not taking into account in the controller design, it may lead to significant deterioration of the control performance and even cause closed-loop instability. Such a performance/stability degradation is usually due to the so called *windup* phenomenon. Such a phenomenon is typically exhibited by dynamic controllers with slow or unstable modes.

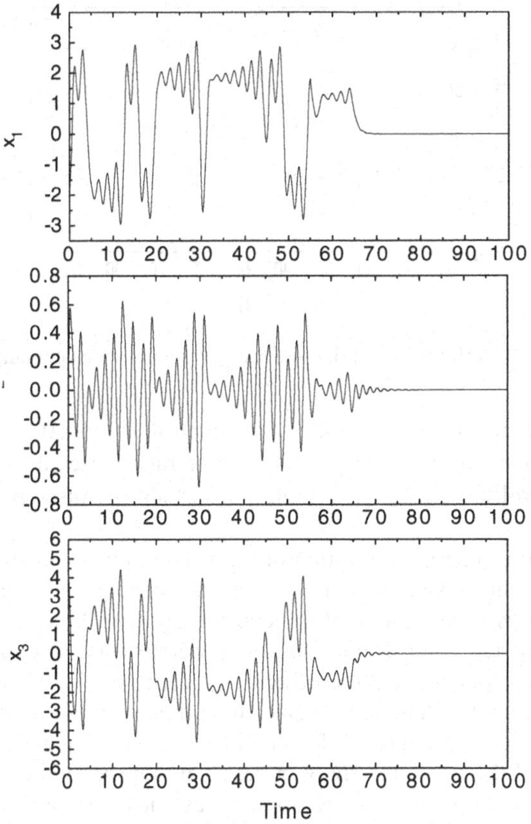

Fig. 2.16. Chaos suppression for the Chua's oscillator, L = 80

Note the controller (2.23),(2.2.4) involves integral actions (i.e., it has a PI structure in conjunction with a correction term of the quadratic integral action), which are respectively given by: a) the difference between the output and the desired value with proportional gain; and b) the integral action induced by the term $\hat{\eta}_{NL} = g_2\int(x_4(t) - \hat{x}_4(t))dt$ to estimate the uncertain terms. Thus, undesired dynamic effects such as the so-called *peaking phenomena* can be induced [52]. Such a phenomenon can produce large overshoots which, under input constraints, can lead to saturation of the control input. In such a situation, the feedback is broken and the plant behaves as an open-loop with a constant input, allowing possible degradation of the close-loop performance. Moreover, due to the integral action of the control law (7), the saturation of the control input can induce undesirable effects such as a large overshoot and settling time [55]. Such a phenomenon is known as *reset windup*. From the implementation point of view, reset windup appears due to the integral action. Since the controller does not have knowledge of the control input saturation, it continues integrating even if the feedback is broken. The saturation is a nonlinear function between the computed

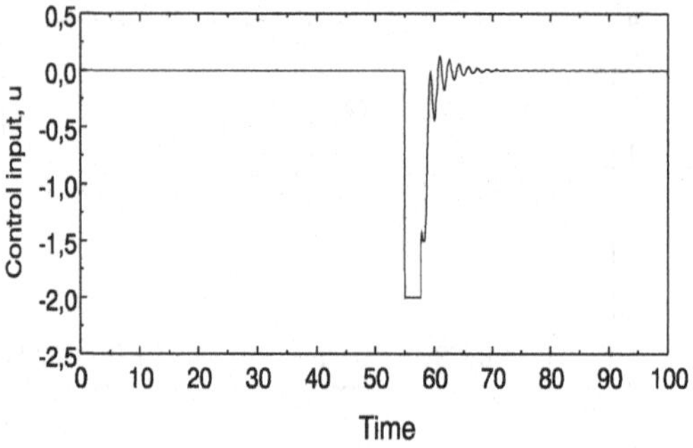

Fig. 2.17. Performance of the saturated version of control input, L = 80

computed control signal (CCS) and the actual control action (ACA) and induces deterioration of the nominal performance. The deterioration rises because the state variables in the control law (2.23) is "winding up" when the discrepancy between CSS and ACA.

According to the general description of the windup phenomenon provided in [56], the control signal can be saturated due to large step changes in reference even when a static control law is used. Note that the control law (2.23) has an internal feedback between the computed control input and the observer used as dynamic estimator of uncertainties η. This means that the observer performance is directly affected by the computed control input. There are schemes to compensate the wind up phenomenon. Recently, an anti-windup approach for nonlinear systems has been proposed as well [57]. The effect of the input overshoot in the chaos suppression of the chuás circuit s is shown in Figure 2.16 and 2.17. Saturation was chosen such that the control command is into the interval $-2.0 \le u \le 2.0$ and L = 80. Here saturation represents the physical bounds of the circuit (voltage and current limits). Note that, when controller in turned on, the control command goes to saturation at -2.0, during such a lapse closed-loop broken because constant input enters to chaotic system.

Exercise 2.23. *Anti-windup in chaos suppression.* The problem of reset windup in the controller (2.23), (2.24) can be treated in a natural way by feeding back the constrained control input (D_{sat}) to the observer. In this way, the controller (2.23),(2.24) can be rewritten in the following form:

$$\begin{aligned} \dot{\hat{z}}_1 &= \hat{\eta} + u_{SAT} + L\kappa_1(z_1 - \hat{z}_1) \\ \dot{\hat{\eta}} &= L^2\kappa_2(z_1 - \hat{z}_1) \end{aligned} \tag{2.25}$$

where u_{SAT} is given by

$$u_{SAT} = \begin{cases} \overline{u}; \ u \ge \overline{u} \\ -\hat{\eta} + \boldsymbol{K}_1(z_1 - \hat{z}_1); \ \underline{u} < u < \overline{u} \\ \underline{u}; \ u \le \underline{u} \end{cases} \tag{2.26}$$

and $\bar{\boldsymbol{u}}$, $\underline{u}$ are the upper and lower bounds. Show that, if the conditions under bounds $\bar{\boldsymbol{u}} = 2.0$ and $\underline{u} = -2.0$, initial conditions $(x_1(0),x_2(0),x_3(0)) = (-2.0,0.02,4.0)$ and $(\hat{\mathbf{z}}_1(0),$

$\hat{\eta}(0)) = (1,15)$ and control parameters as follows: $L = 80$, $k_1 = 1.0$ and $(\kappa_1,\kappa_2) = (2.0,1.0)$, do the anti-windup scheme (2.25),(2.26) improve the controller performance ? ◦

2.5 Suppressing Chaos in Second-Order Systems

The classical PI and PII^2 feedback for chaos control was discussed and illustrated in previous sections. Such a control has been illustrated and design conditions for stabilizing at origin were established as well. Result shows that if relative degree is equal to one, the PI with dynamical estimation (PII^2) is obtained from the procedure proposed in Section 2.4. However, if condition for relative degree equal to 1 is not satisfied, the resulting controller is more complex than PI or PII^2 feedback. In order to illustrate the chaos suppression with relative degree larger than one, we shall show two examples in the following paragraphs. The examples correspond to second-order driven oscillators. Moreover, here we extend the results on stabilization at origin. In fact, chaos suppression problem consists in leading the chaotic trajectories to periodic orbits are equilibrium (fixed) points.

Several physical systems can be modelled by means of second order nonautonomous ordinary differential equations. It is well known that under certain conditions, this class of systems can show a very complicated behavior. For instance, mechanical systems with friction can display, under forcing conditions, a behavior which includes noninvertible erratic trajectories [58]. Chaotic behavior has been found in a superconductor when a magnet is excited periodically in lateral motion [59] and chaos is induced by hysteresis forces and stiffness in a levitated magnet near to the surface of a high T_c superconductor (HTS) [60][61] . However, in many practical situations it may be desirable that a given system originally undergoing complicated behavior should be forced to displays regular motions (*i.e.*, suppression of chaos). For instance, it could be desirable to induce regular dynamics in mechanical oscillators to avoid errors (as in the case of precise position mechanisms) leaded by external vibrations and magnetic fields. Also, it should be desirable regulate the nonlinear dynamics in the high speed rotor supported by superconductor which might involve lateral and vertical nonlinear forces [62]. On the other hand, there are three main types of strategies to design control schemes in nonlinear systems, these are briefly discussed in the following paragraphs.

Continuous feedback controls have been used by several authors in order to suppress complicated nonlinear behavior [9],[10]. Pyragas [9] suggested a continuous control scheme where the chaos suppression is allowed by means of an experimentally adjustable parameter. However, Pyragas' control scheme requires exact knowledge of the system model which is not a realistic situation. A modal feedback control is developed in [10] which directly alters the Lyapunov exponents in linear systems where the model is exactly known. The fact that the strategy proposed in [10] requires the exact knowledge of the system model is a disadvantage. Another control scheme to induce regular dynamics was proposed by Chen and Chou [63] which is based on continuous feedback. Chen-Chou's strategy makes use of a coordinates change to transform the original system in a controllable system, however, this coordinates change requires of the knowledge of all states of the original system. Thus, the control scheme only can be used if all states are available via measurements (state feedback control).

Recently, more robust control strategies, borrowed from conventional engineering methods of control theory, have been considered. These methods include linear feedback control [64], adaptive control techniques [65], discrete-time strategies of torque balance [58] and linearizing robust control [66][20]. A central issue when applying such control techniques is to obtain an acceptable control performance in spite of fluctuations in parameters due to external effects and noise disturbances. In other words, robust stability margins against uncertainties in the model and perturbations (see for instance [60],[61],[20]). Linear control methods are based on Taylor linearization of the system around a prescribed reference trajectory are excellent control schemes if the principal dynamical characteristic of the system to be controlled can be retained in the linear terms. Although there exist many results about stabilization and trajectory tracking of linear systems [59], the application of linear feedback control to suppress chaotic signals might not be a good idea since chaos has nonlinear nature and the principal dynamical characteristics can be lost in the Taylor linearization of the system. Adaptive techniques have the drawback that linearity in the parameters to be estimated is assumed [60]. In addition, to handle model uncertainties with adaptive techniques is not an easy task. On the other hand, discrete-time strategies suggest the estimation of model uncertainties and external forces as a *new state* [53]. However, discrete-time strategy has the problem that measurements noise has adverse effects on the performance of the controller because it is an unstable operation.

First let us discuss a general mathematical model for the levitation. The model is represented by a nonautonomous ordinary differential equation which include the dynamic magnetic force. Consider a magnet supported by superconducting system which can be represented by a second order differential equation with a nonlinear term which involves hysteresis and periodic external excitation force [55]. Without loss of generality, one can consider that the model of the levitation system is modelled by the following dimensionless equation

$$\ddot{x} + \delta\dot{x} + x - \alpha(x,\dot{x},t) = \tau_e + \tau \tag{2.27}$$

where x is the displacement of the magnet from the HTS surface, δ represents a mechanical damping coefficient, $\alpha(x,x,t)$ is the force between the HTS and the magnet, τ is the control force, $\tau_e = A_1\cos(\omega t)$ is the external excitation force, A_1 and ω denote the amplitude and driving frequency, respectively. By defining $x_1 = x$ and $x_2 = \dot{x}$ the equation (2.27) can be written as the following system $\dot{x}_1 = x_2$ and $\dot{x}_2 = \alpha(x_1,x_2,t) - \delta x_2 - x_1 + \tau_e(t) + \tau$, which is equivalent to (2.5) if we define $f(x) = [x_2\ \alpha(x_1,x_2,t) - \delta x_2 - x_1]^T$, $d(t) = [0\ \tau_e(t)]^T$, $u = \tau$ and $g(x) = [0\ 1]^T$. It must be pointed out that the superconducting levitation system can exhibit chaotic behavior in the presence of hysteresis if the force-displacement relation is excited periodically [55].

A lot of effort has been devoted to the modelling of the magnetic force $\alpha(x_1,x_2,t)$. However, it has been demonstrated that the magnetic induction in HTS is a nonlinear phenomena with a number of nontrivial peculiarities (even in the one-dimensional case) which can be attributed to a conjunction of two factors [67]: (a) the impossibility of linearization of the diffusion of the magnetic flux in a soft superconductor and (b) the vector nature of the magnetic induction. In other words, when an alternating external magnetic field is applied on a superconductor surface, the type-II HTS plays a role of a unique nonlinear element, a damper, in such way that the magnetic-field oscillations penetrate into the interior of the specimen only to depth l of the order of $\omega^{-1/2}$ (see [62] and references therein). Moreover, since the magnetization is not uniform throughout the superconductor, a large calculation effort and many magnetization measurements would be required for an accurate prediction of the magnetization behavior of the superconductor as the applied field varies [68]. Besides, it has been found that magnetic stiffness and levitation force between a magnet and a superconductor surface vary from several major types of HTS. Additionally, the magnet-superconductor force exhibits a repeatable hysteresis loop as the position of the magnet [63]. Thus, it is clear that $\alpha(x,x,\mathrm{t})$ can not be available for feedback from measurements.

So, as an illustrative example, we state the control problem as follows: *Given a target continuous-time trajectory r(t), design a control feedback such that the solution x(t) of the system (2.27) tracks r(t) despite lack in the magnet-superconductor force model and unmeasured external disturbances.*

In what follows, we describe a strategy to control the levitated magnet vertical motion of the superconducting bearing and discuss it in physical terms towards inductive extension of the PI and PII^2 feedback. Thus, the proposed strategy is a feedback control which reduces the effect of unmeasured external disturbances and model uncertainties of the levitation system. The result are directly related with Section 2.4. The above control problem is addressed under the assumptions in previous sections, which can be worded for the levitation system as: *1)* only the position x_1 is available for feedback and *2)* the magnet-superconductor magnetic force $\alpha(x_1,x_2,\mathrm{t})$ and the external force $\tau_e(\mathrm{t})$ are unknown. Some comments regarding the physical meaning of the assumptions are in order. In most practical situations, it is only possible to measure one observable which is most commonly position. In fact, in order to measure velocity, position measurements are required. On the other hand, $\alpha(x_1,x_2,\mathrm{t})$ and $\tau_e(\mathrm{t})$ can not be exactly known since imperfections in the model (for instance, the concerning to the discussion of the above section). One potential drawback of linearizing control techniques is that they rely on exact cancellation of nonlinear terms in order to obtain a linear behavior of the system to be controlled [58],[36]. However, at same manner that previous section, we use an intermediate supposition toward the design of the robust controller which will depend only on position and estimates of uncertainties. The intermediate supposition is the following: *Let us suppose that* $\alpha(x,\dot{x},\mathrm{t})$ *and* $\tau_e(\mathrm{t})$ *are exactly known.* Let $\chi(\mathrm{t}) \in \mathbb{R}^2$ be a prescribed position trajectory. If the terms $\alpha(x_1,x_2,\mathrm{t})$ and $\tau_e(\mathrm{t})$ were known, linearizing feedback control is given by [36]

$$\begin{aligned} \tau_1(x) &= -\alpha(x_1,x_2,t) + \delta\dot{x} + x - \tau_e + V(x_1,x_2,r) \\ V(x_1,x_2,\chi) &= \ddot{r} - g_1(x_2 - \dot{r}) - g_2(x_1 - r) \end{aligned} \tag{2.28}$$

where the control parameters g_1 and g_2 are chosen in such way that the polynomial $P_2(\lambda) = \lambda^2 + \kappa_1\lambda + \kappa_2$ has roots in the open left-half complex plane ($P_2(\lambda)$ is so-called Hurwitz). The linearizing control law (2.28) would asymptotically steer the system trajectories to a behavior with $x_1(\mathrm{t}) = \dot{r}(\mathrm{t})$ and $x_2(\mathrm{t}) = \mathrm{r}(\mathrm{t})$. The control law (2.24) can be called *Ideal Feedback Control* (IFC) because its physical implementation requires a perfect knowledge of the dynamics of the levitation force and velocity measurements. The IFC (2.28) has a nice feature, it induces a linear dynamical behavior to the error tracking $e = x - r$. In fact, for (2.28) system (2.27) can be written as $\ddot{e} + \kappa_1\dot{e} + \kappa_2 e = 0$ [36].

However, if the velocity is not available for feedback, the linearizing control law (2.28) can not be directly implemented. In fact, by the assumption S.1, an estimate of x_2 is required. In such case, by following ideas in Sect. 2.4, we have that, defining $\eta(t) = \alpha(x_1(t),x_2(t),t) + \tau_e(t) - \delta x_2 - x_1$. The system (2.27) can be rewritten as an externally dynamically equivalent system which is represented by the following extended system

$$\begin{aligned} \dot{x}_1 &= x_2 \\ \dot{x}_2 &= \eta + \tau \\ \dot{\eta} &= \Gamma(x,\eta,t) \\ &= x_2\partial_1\alpha + (\eta + \tau)\partial_2\alpha + \partial_t\alpha + \dot{\tau}_e \end{aligned} \tag{2.29}$$

where $\partial_1\alpha = \partial\alpha(x_1,x_2,t)/\partial x_1$, $\partial_2\alpha = \partial\alpha(x_1,x_2,t)/\partial x_2$ and $\partial_t\alpha = \partial\alpha(x_1,x_2,t)/\partial t$. The manifold $\Phi = \{(x,\eta) \in \mathbb{R}^3 : \eta - \alpha(x_1,x_2,t) - \tau_e(t) + \delta x_2 + x_1 = 0\}$ is invariant under the trajectories of (2.29). That is, $d\Phi/dt = 0$ along the trajectories generated by the vector field in (2.29). Moreover, if the initial condition $(x(0),\eta(0)) \in \Phi$, the system (2.29) has the same solutions as the system (2.27), *for all control inputs* $\tau(t)$. Consequently, it is equivalent to design a controller for system (2.29) as for the system (2.27). The extended system (2.29) reproduces the dynamical behavior of the mechanical systems (2.27)

Then, in coordinates $(\hat{x},\eta)$, the IFC becomes:

$$\tau(x,\eta) = -\eta + V(x_1,x_2,r) \tag{2.30}$$

with $V(x_1,x_2,r)$ as in (2.28). In order to implement (2.30), estimates of the unmeasured velocity $x_2(t)$ and of the unknown 'state' η are required. Given measurements of the position $x_1(t)$, it is easy to see that the system (2.29) is observable [49]. That is, the dynamic of the states (x_2,η) can be reconstructed from the (measured) dynamics of $x_1(t)$. However, time-derivation of $x_1(t)$ can not be easily implemented because derivation is an unstable operation (high sensitivity to noise measurements). An alternative to direct derivation is the use of observers [49] and [51]. As in the case of velocity estimation, the following state estimator is proposed

$$\begin{aligned} \dot{\hat{x}}_1 &= \hat{x}_2 + L\kappa_1^*(x_1 - \hat{x}_1) \\ \dot{\hat{x}}_2 &= \hat{\eta} + \tau + L^2\kappa_2^*(x_1 - \hat{x}_1) \\ \dot{\hat{\eta}} &= L^3\kappa_3^*(x_1 - \hat{x}_1) \end{aligned} \tag{2.31}$$

where $(\hat{x}_1, \hat{x}_2, \hat{\eta})$ are estimates of (x_1, x_2, η), respectively, $L > 0$ (estimator gain) is an adjustable parameter. Notice that the uncertain term $\Gamma(x,\eta,t)$ was not considered in the design of the estimator (2.31). The constants $[\kappa^*1,\kappa^*2,\kappa^*3]$ are chosen in such way that the companion matrix

$$A = \begin{bmatrix} -\kappa_1^* & 1 & 0 \\ -\kappa_2^* & 0 & 1 \\ -\kappa_3^* & 0 & 0 \end{bmatrix} \tag{2.32}$$

has all its eigenvalues in the open left-half complex plane. By defining the estimation error as $e_1 = L^2(x_1 - \hat{x}_1)$, $e_2 = L(x_2 - \hat{x}_2)$ and $e_3 = (\eta - \hat{\eta})$, the error dynamics can be written as follows

$$\dot{e} = LA(\beta)e + \Phi(x,\eta,N(L),e,t) \tag{2.33}$$

where $N(L) = diag(L^{-2}, L^{-1},1)$, and $\Phi(x,N(L)\varepsilon,t)$ is a nonlinear function of its arguments. In this way, $\hat{x}_2(t)$ becomes an estimate of the velocity and $\hat{\eta}(t)$ is an estimate of the uncertain terms $\alpha(x_1,x_2,t) + \tau_e(t) - \delta x_2 - x_1$. Using the estimates the implemented feedback controller is given by

$$\tau(t) = -\hat{\eta} + V(\hat{x},\hat{\eta},r) \tag{2.34}$$

It is possible to prove that for $L > 0$ sufficiently large, $|e(t)| \to O(L^{-1})$ as $t \to \infty$, and the position $x_1(t)$ tracks the reference signal $\omega(t)$ with an error $O(L^{-1})$ when initial conditions $x(0)$, $\dot{x}(0)$ are contained in an arbitrary compact set (*semiglobal practical stabilization*) [20].

2.5.1 Sketching the Stability Analysis

Without loss of generality, assume that $r(t) = 0$ for all $t \geq 0$ (this means that $r(t) = 0$). Let us now consider the following nonlinear system

$$\begin{aligned} \dot{x}_1 &= x_2 \\ \dot{x}_2 &= \eta + \tau \\ \dot{\eta} &= \Gamma(x,\eta,t) \\ &= x_2\partial_1\alpha + (\eta+\tau)\partial_2\alpha + \partial_t\alpha + \dot{\tau}_e \end{aligned} \tag{2.35}$$

where $\partial_1\alpha = \partial\alpha(x_1,x_2,t)/\partial x_1$, $\partial_2\alpha = \partial\alpha(x_1,x_2,t)/\partial x_2$ and $\partial_t\alpha = \partial\alpha(x_1,x_2,t)/\partial t$.

Proposition 2.24. The nonlinear system (2.35) is externally equivalent to the nonlinear system (2.27).

Proof. Let us define $\Theta(x_1,x_2,u,t;\pi) = \alpha(x_1(t),x_2(t),t) + \tau_e(t) - \delta x_2(t) - x_1$ and $\Psi(x_1,x_2,\eta,\nu,u,t;\pi) = \eta - \Theta(x_1,x_2,u,t;\pi)$. Since $\alpha(x_1(t),x_2(t),t)$, $\tau_e(t)$ and $\delta x_2(t)$ are smooth functions hence $\Theta(x_1,x_2,u,t;\pi)$ and $u = \{\Gamma(x_1,x_2,u,t;\pi) - k_d(x_2 - x_d) - k_p(x_1 - x_d)\}$ are smooth, consequently $\Psi(x_1,x_2,\eta,\nu,u,t;\pi) := \eta(t) - \alpha(x_1(t),x_2(t),t) - \tau_e(t) + \delta x_2(t) + x_1 = 0$ is a *first-integral* of the nonlinear system (2.33). Since $\Psi(x_1,x_2,\eta,\nu,u,t;\pi) = 0$ is a *first-integral* of (2.35) hence one has that along the trajectories of system (2.35) $d\Psi(z,\eta,\nu,u)/dt = 0$ for all $t \geq 0$. This implies that the manifold $\Psi(x_1,x_2,\eta,\nu,u,t;\pi) = \{(x,\eta) \in \mathbb{R}^3\colon \eta - \Theta(x_1,x_2,u,t;\pi) = 0\}$ is invariant under the trajectories of (2.35). In other words, the state representation (2.35) is externally equivalent to the system (2.27), as long as initial conditions satisfy $\Psi(x_1(0),x_2(0),\eta(0),u(0),0;\pi) = 0$. □

Remark 2.25. As in Sect. 2.4, the augmented state variable $\eta(t)$ provides the dynamics of the uncertain function $\Theta(x_1,x_2,u,t;\pi)$ which involves model and parametric uncertainties.

Remark 2.26. One can take eq. (2.35) and the algebraic constrain $\Psi(x_1,x_2,\eta,\nu,u,t;\pi) = 0$ as a state representation of (2.27). Hence, whenever be necessary one can use the fact that $\eta = \Theta(x_1,x_2,u,t;\pi)$ as a tool to avoid the inclusion of the state η in the stability analysis of the resulting closed-loop system.

Proposition 2.27. Since the systems (2.27) and (2.35) are externally dynamically equivalents, one can design the next linearizing control law

$$u(t) = -\eta - \kappa_2 x_1 - \kappa_1 x_2 \tag{2.36}$$

under control law (2.36) the states (x_1,x_2) of system (2.36) will converge asymptotically to the pair (r,r), respectively.

Proof. Convergence of (x_1,x_2) to (r,r) follows from the fact that $\Psi(x_1,x_2,\eta,v,u,t;\pi) = \eta - \Xi(x_1,x_2,u,t;\pi)$ is a first integral of the closed-loop system. Then, we have that $\eta(t) - f(x_1(t),t;\pi) - \delta x_2(t) = 0$, or equivalently $\eta(t) - \kappa_2 x_2 - \kappa_1 x_1 = 0$, this shows that $\eta(t)$ remains bounded for all $t \geq 0$. □

Since the system (2.35) is observable and the *new state*, by following ideas o} in Sect. 2.4, $\eta(t)$ can be reconstructed from system output $y = x_1$, consequently the dynamics of the uncertain function $\Theta(x_1,x_2,u,t;\pi)$ can be reconstructed via the state $\eta(t)$. However, in order to design a state observer for nonlinear system (2.33), it must be known the term $\Theta(x_1,x_2,u,t;\pi)$. To avoid this problem, we propose the following state estimator

$$\begin{aligned} \dot{\hat{x}}_1 &= \hat{x}_2 + L\kappa_1^*(x_1 - \hat{x}_1) \\ \dot{\hat{x}}_2 &= \hat{\eta} + \tau + L^2\kappa_2^*(x_{1\_}\hat{x}_{11}) \\ \dot{\hat{\eta}} &= L^3\kappa_3^*(x_1 - \hat{x}_1) \end{aligned} \tag{2.37}$$

where κ_1^*, κ_2^* and κ_3^* are chosen such that the polynomial $s^3 + \kappa^*1s^2 + \kappa^*2s + \kappa^*3 = 0$ is Hurwitz and $L > 0$. Thus, the linearizing control law becomes

$$u(t) = -\hat{\eta} - \kappa_2\hat{x}_1 - \kappa_1\hat{x}_2 \tag{2.38}$$

Proposition 2.28. For L sufficiently large, the closed-loop system (2.37),(2.38) is asymptotically stable.

Proof. Let us define $e_1 := L^2(x_1 - \hat{x}_1)$, $e_2 := L(x_2 - \hat{x}_2)$ and $e_3 := \eta - \hat{\eta}$. Thus, the closed-loop system is represented by

$$\begin{aligned} \dot{x}_1 &= x_2 \\ \dot{x}_2 &= -\eta + q\cos(\omega t) + u(t) \\ \dot{\eta} &= \Gamma(x,\eta,u,t) \\ \dot{e} &= LA(\kappa)e + \Phi(x,\eta,e;L) \end{aligned} \tag{2.39}$$

where $\Phi(x,\eta,e;L) = [0, \Gamma(x_1,x_2,u,t)]^T$, $e = [e_1,e_2,e_3]^T$ and the companion matrix $A(\kappa)$ is given by

$$A(\beta) = \begin{bmatrix} -\kappa_1^* & 1 & 0 \\ -\kappa_2^* & 0 & 1 \\ -\kappa_3^* & 0 & 0 \end{bmatrix} \tag{2.40}$$

which is, by assumption, Hurwitz. On the other hand, we can use the constraint $\eta = \alpha(x_1(t),x_2(t),t) + \tau_e(t) - \delta x_2(t) - x_1$ to eliminate the state η of the equations (2.39).

In fact, using the above constraint, the feedback controller (2.36) can be rewriten as the linearizing control law (2.35), (2.36). Then, if $\Psi(x_1(0),x_2(0),\eta(0),u(0),0;\pi) = 0$, $\eta(t) = \alpha(x_1(t),x_2(t),t) + \tau_e(t) - \delta x_2(t) - x_1$ and the controller (2.36) behaves as linearizing control law (2.26). □

Corollary 2.29. When the initial conditions are contained in an arbitrary compact set, the saturation function of the controller (2.36) ensures that $lim \mid e(t) \mid \rightarrow O\ (L^{-1})$ as $t \rightarrow \infty$ (semiglobal practical convergence).

Furthermore, as $(\hat{x}(t),\hat{\eta}(t)) \in \mathbb{R}^2 \times \mathbb{R}$ approaches $(x(t),\eta(t))$, the feedback controller (2.32) approaches the IFC (2.24). Besides, note that the controller (2.29),(2.32) does not require knowledge respect the function of the magnet-superconductor force. In this way, the proposed controller can be implemented experimentally.

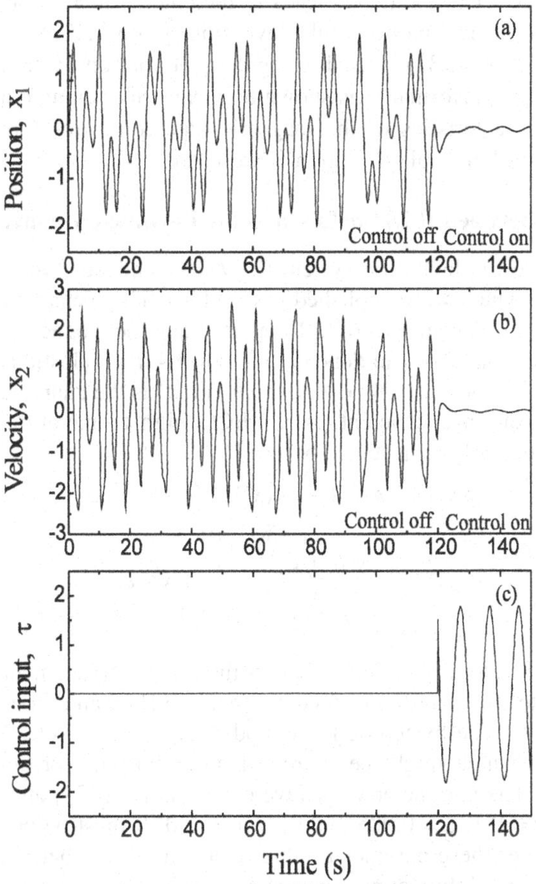

Fig. 2.18. Chaos suppression of the van der Pol equation. (a) Position $x_1(t)$, (b) Velocity, $x_2(t)$ and (c) Control command $\tau(t)$.

The ability of the controller (2.37),(2.38) to suppress chaos has been demonstrated in [20]. However, we have controlled the chaotic behavior of the van der Pol equation to show that results in previous sections can be extended for the case of relative degree larger than 1. Thus, the nonlinear driven oscillator to be controlled becomes

$$\ddot{x} + \mu(x^2 - 1)\dot{x} + x^3 = \alpha \cos(\omega t) \tag{2.41}$$

where $\mu = 0.1$, $\alpha = 1.75$ and $\omega = 0.667$. Figure 2.18 shows the stabilization of eq. (2.41) at equilibrium point (0,0). The control constants k_1 and k_2 were chosen such that the polynomial $P_2(\lambda) = \lambda^2 + k_1\lambda + k_2 = 0$ has all the roots located at -1.0. The values of κ_i^*'s were chosen such that the companion matrix (2.40) has all its eigenvalues located at -1.0.

The parameter estimation value was chosen as $L = 50.0$. The controller was activated at $t \geq 120$ (this means that $\tau = 0$ for $t < 120$). It is clear that the controller (2.29),(2.32) allows the suppression of the chaotic behavior of the van der Pol oscillator. Since the controller (2.29),(2.32) leads to the chaos suppression, hence it can be used to regulate the vertical motion of the levitation system (2.27).

The controller (2.29),(2.32) is used to regulate the motion of the levitation system (2.27). The controller performance is shown by means of computer simulations. First, we recall some characteristics of the magnetic-superconductor force, thereafter, an example of the controller implementation is presented.

2.5.2 The Force between HTS Surface and the Levitated Magnet

In order to simulate the levitation system (2.27), we choose some characteristics of the magnetic force which were published in [59],[60]. These characteristics retain the behaviour of $\alpha(x, \dot{x}, t)$. The authors of [59] reported that the magnetic characteristics of an antiferromagnetic material can depend on modified hypo-elasticity function whose dynamic is $\dot{\alpha}(x, \dot{x}) = \gamma[\alpha(x,\dot{x},t) - \varphi(x, \dot{x})]$ where $f(x, \dot{x})$ is the characteristic of hysteresis and γ is a relaxation coefficient. In a YBCO high temperature superconductor, $\varphi(x, \dot{x})$ can be particularly represented by

$$\varphi(x,\dot{x}) = \varphi_1(x)[1 + \varphi_2(x)], \quad \varphi_1(x) = Fe^{-x}$$

$$\varphi_2(\dot{x}) = \begin{cases} -\mu_1 - \dot{x}, & \varepsilon \leq \dot{x} \\ \dfrac{-\dot{x}(\mu_1 + \mu_2)}{2\varepsilon} & -\varepsilon \leq \dot{x} \leq \varepsilon \\ \mu_2 & \dot{x} < -\varepsilon \end{cases} \tag{2.42}$$

where the exponential term $f_1(x) = Fe^{-x}$ shows the force-displacement relation without hysteresis, F denotes the maximum force between the HTS and the magnet, μ_1 and μ_2 are typical constants. The hysteretic force model (2.41) gives to (2.27) the additional degree of freedom which might be source of quasi-periodic (or chaotic) orbits and uncertainties [60]. The parameter values were chosen as in [59] and they are the following: $F = 0.3$, $\gamma = 0.1$, $\varepsilon = 0.005$ and $\mu_1 = \mu_2 = 1.0$. It must be pointed out that the equation (2.27) under these parameter values can exhibit non-periodic orbits. In fact, the levitation system exhibits almost-periodic behavior. In the Figure 2.19, the phase

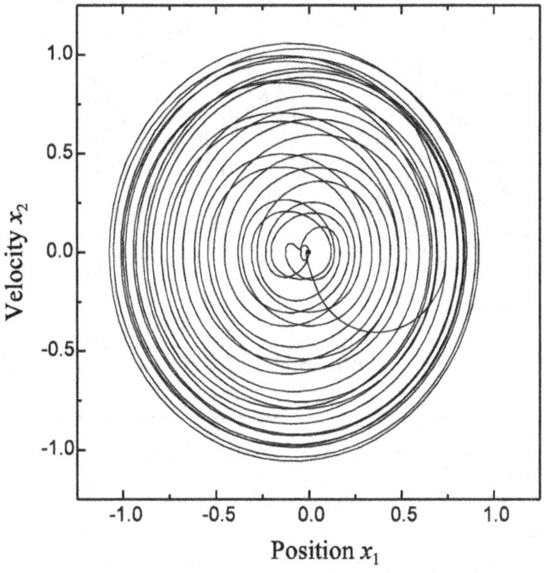

Fig. 2.19. Phase portrait of the levitation system without control actions, $\tau = 0$ for all $t \geq 0$

portrait of the system equivalent to (2.27) (without control actions, $\tau = 0$ for all $t \geq 0$) is presented. Moreover, there exist experimental evidence that superconducting bearing can display chaos.

Assume that the magnet-superconductor force is given by equation (2.42). Also, let us suppose that the target trajectory is $r(t) = 0$, hence $\dot{r}(t) = 0$. The control parameter $k_1 = k_2 = 1.0$ were chose, in such way the Hurwitz polynomial $P_2(\lambda)$ has all the roots in (-1.0,-1.0). In addition, the estimation parameter L = 50. At time t = 160 the control is activated, (this means that at time t < 150 the control command $\tau = 0$). Figure 2.20a shows the position trajectory before and after that the control activation. In the Figure 2.20b, the control command is displayed.

It is easy to see that the system is stabilized despite the controller (2.37),(2.38) does not know the functionality of the magnet-superconductor force, actual values of the velocity nor external excitation force. However, the cost for robustness is a large value of the estimation parameter which yields the saturation of the control command. On the other hand, the oscillation of the control command $\tau(t)$ (at t > 160) is due to the compensation of the external excitation force. The controller (2.37),(2.38) comprises a dynamic uncertainties compensator (a Luenberger state estimator) and a linearizing-like controller. The uncertain terms were lumped in a time variable which can be interpreted as a new state. The dynamic of the new state is reconstructible from position measurements, in this way the information required to control the vertical motion of the magnet is provided.

Computer simulations show that by means of the proposed control scheme a target position trajectory can be tracked in spite of lack in: (a) the magnet-superconductor force model, (b) external excitation force knowledge and, (c) actual value of the

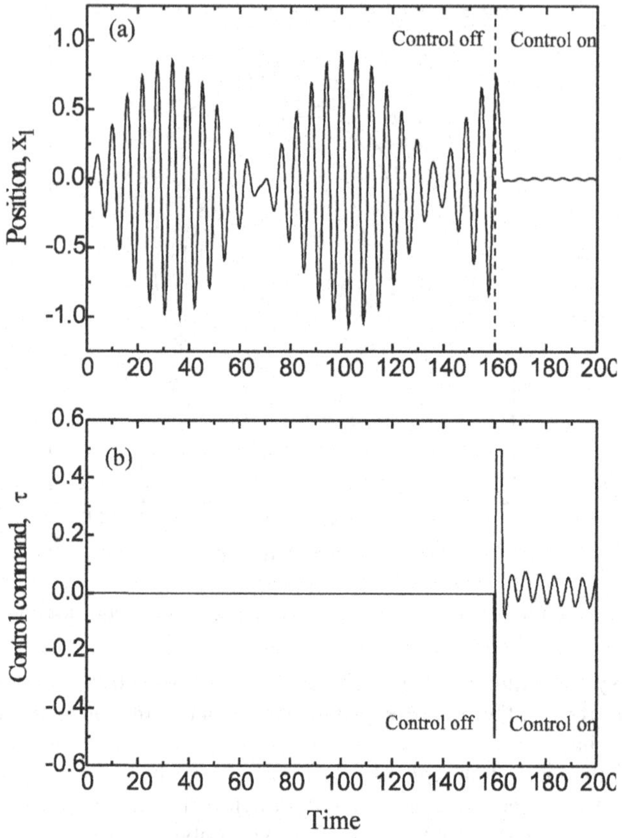

Fig. 2.20. System response for control gain $g_1 = g_2 = 1.0$ and estimation parameter $L = 50$. (a) Position $x_1(t)$ and (b) Control command $\tau(t)$.

velocity. However, the cost for robustness is a large estimation parameter value. A large value of the estimation parameter is a disadvantage, because overshoot (saturating control actions) can be induced. To avoid this drawback, for example, a nominal model of the magnet-superconductor force can be provided in such way that the modelling errors be lumped with the external excitation forces and velocity. Hence, one can expect a low value of the estimation parameter.

Exercise 2.30. *Computing the convergence time in chaos suppression.* Consider that the second-order system (2.27) is given by the Duffing equation (see [1] for details). Assume that (i) the Duffing equation has been rewritten in the form of (2.35) and (ii) the state feedback (2.30) takes the form $\tau = u = \ddot{r} - \eta - \frac{1}{2}\mathbf{B}^T\mathbf{N}^{-1}e$, where $\triangleq (e_1,e_2)^T = ((x_1 - r),(x_2 - r))$ stands for the control error, r denotes the reference signal, $B = [0, 1]$ is a constant vector and N^{-1} is the inverse matrix whose entries are given by

$$N_{i,j} = \frac{(-1)^{i+j}\left(a/b\right)^{5-i-j}(4-i-j)!}{(2-i)!(2-j)!\left[(a+1)\ldots(a+5-i-j)\right]}$$

where $a > 1$ and $b > 0$ are given constants, and $i,j = 1,2$. Prove that the error trajectories $e_1(t)$ and $e_2(t)$ converge to zero at a finite time $T = (a/b)\theta^{1/a}(e(0))$, where $e(0)$ stands for any initial condition at an open set $\Omega \subset \mathbb{R}^2$. **Hint:** The authors recommend to review [69], where a similar result can be found. ○

2.6 Summary

Here a robust approach for suppressing chaos. The chapter has started with frequency domain to, in an inductive manner, land in a robust approach for a geometrical feedback and is mainly focussed on second- and third-dimensional systems with relative degree equal to 1 or 2. Some control issues are commented and exercises show extensions toward higher order systems whose relative degree can be larger than 1 or 2 and complementary control aspects (e.g., anti-reset windup). The chaos control problem in context of suppression has been widely studied and interested reader can find a large number of open literature. The following references are sample of theoretical experimental issues on chaos suppression and its relation with the results in the this chapter. Following same inductive structure, next chapters will address the synchronization of chaotic uni-directionally coupled systems and exercises and examples show extensions for different interconnection for studying synchronization.

References

[1] Gukenheimer, J., Holmes, P.: Nonlinear oscillations, dynamical systems and bifurcations of vector fields. Springer, N.Y (1983)

[2] Luna-Rivera, M., Femat, R.: A study in frequency of controlled chaotic systems (in Spanish). Rev. Mex. Fis. 46, 429 (2000)

[3] Ho, M.C., Ko, J.Y., Yang, T.H., Chen, J.L.: A generic input-output analysis of zero-dispersion noninear resonance. Europhys. Letts. 48, 603 (1999)

[4] Lima, R., Pattini, M.: Suppression of chaos by resonant parametric perturbation. Phys. Rev. A 41, 726 (1990)

[5] Ott, E.: Chaos in dynamical systems. Cambridge University Press, Cambridge (1992)

[6] Anischenko, V.S.: Dynamical chaos: models and experiments. World Scientific, Singapore (1995)

[7] Piserchik, A.N., Corbalán, R.: Stochastic resonance in chaotic laser. Phys. Rev. E 58, 2697 (1998)

[8] Aguirre, L.A., Billings, S.A.: Model reference control of regular and chaotic dynamics in the Duffing-Ueda oscillator. IEEE Trans. Circ. & Syst. I 41, 477 (1994)

[9] Pyragas, K.: Continuous control of chaos by self-controlling feedback. Phys. Lett. A 170, 421 (1992)

[10] Wiesel, W.E.: Modal feedback control on chaotic trajectories. Phys. Rev. E. 49, 1990 (1994)

[11] Nijmeijer, H., Berghuis, H.: On Lyapunov control of the Duffing equation. IEEE Trans. Circ. & Syst. I 42, 473 (1995)

[12] Aguirre, L.A., Billings, S.A.: Closed-loop suppression of chaos in nonlinear driven oscillators. J. Nonlinear Sci. 5, 189 (1995)

[13] Femat, R., Capistrán-Tobías, J., Solís-Perales, G.: Laplace domain controllers for chaos control. Phys. Lett. A 252, 27 (1999)

[14] Wei, W.W.-S.: Time series analysis. Addison-Wesley, USA (1990)

[15] Lorenz, E.N.: Deterministic nonperiodic flow. J. Atmos. Sci. 20, 130 (1963)

[16] Haken, H.: Synergetics an introduction. Springer, Berlin (1983)

[17] Arnold, V.I., Afraimovich, V.S., Il'yashenko, Y.S., Shl'nikov, L.P.: Bifurcation theory and catastrophe theory. Springer, Heidelberg (1999)

[18] Halle, K.S., Wu, C.W., Itoh, M., Chua, L.O.: Spread spectrum communication through modulation of chaos. Int. J. of Bifur. and Chaos 3, 469 (1993)

[19] Alvarez-Ramírez, J.: Nonlinear feedback for controlling the Lorenz equation. Phys. Rev. E 50, 2339 (1994)

[20] Femat, R., Alvarez-Ramírez, J., González, J.: A strategy to control chaos in nonlinear driven oscillators with least prior knowledge. Phys. Lett. A 224, 271 (1997)

[21] Mossayebi, F., Qammar, H.K., Hartley, T.T.: Adaptive estimation and synchronization of chaotic systems. Phys. Lett. A 161, 255 (1991)

[22] Kozlov, A.K., Shalfeev, Chua, L.O.: Exact synchronization of mismatched chaotic systems. Int. J. of Bifur. and Chaos 6, 569 (1996)

[23] Alvarez-Ramírez, J., Femat, R., González, J.: A time-delay coordinates strategy to control a class of chaotic oscillators. Phys. Lett. A 211, 41 (1996)

[24] Wu, C.W., Yang, T., Chua, L.O.: On adaptive synchronization and control of nonlinear dynamical systems. Int. J. Bifur. and Chaos 6, 455 (1996)

[25] di Bernardo, M.: An adaptive approach to the control and synchronization of continuous-time chaotic systems. Int. J. Bifur. and Chaos 6, 557 (1996)

[26] Yu, X., Chen, G., Xia, Y., Song, Y., Cao, Z.: An invariant-manifold-based method for chaos control. IEEE Trans. on Circuits and Systems I. 48, 930 (2001)

[27] Rulkov, N.F., Sushchik, M.M.: Robustness of synchronized chaotic oscillators. Int. Jour. of Bifur. and Chaos 7, 625 (1997)

[28] Narendra, K.S., Annaswamy, A.M.: Stable adaptive systems. Prentice-Hall, N.J. (1989)

[29] Ott, E., Grebogi, C., Yorke, J.A.: Controlling chaos. Phys. Rev. Letts., 1196 (1990)

[30] Roy, R., Murphy Jr., T.W., Maier, T.D., Gills, Z., Hunt, E.R.: Dynamical control of a chaotic laser: Experimental stabilization of a globally coupled system. Phys. Rev. Lett. 68, 1259 (1992)

[31] Christini, D.J., Collins, J.J.: Real-time model-independent control of low dimensional chaotic and nonchaotic systems. IEEE Trans. on Circuits and Systems I 44, 1027 (1997)

[32] Femat, R., Alvarez-Ramírez, J.: Synchronization of a class of strictly-different oscillators. Phys. Lett. A 236, 307 (1997)

[33] Coughanouwr, D.R., Koppel: Process systems analysis and control. McGraw-Hill, USA (1965)

[34] D'Azzo, J.J., Houpis, C.H.: Linear control system analysis and design. McGraw-Hill, Tokyo (1975)

[35] Alvarez-Ramírez, J., Femat, R., Barreiro, A.: A PI controller with dynamic estimation. Ind. Chem Eng. Res. 36, 3668 (1997)

[36] Isidori, A.: Nonlinear control systems. Springer, Berlin (1989)

[37] Nijmeijer, H.: Nonlinear dynamical control systems. Springer, N. Y (1990)

[38] Doyle, J.C., Francis, B.A., Tannembaum, A.R.: Feedback control theory. MacMillan Publ. Company, N.Y (1992)

[39] Morari, M., Zafirou, E.: Robust process control. Prentice-Hall, N.J (1989)

[40] Kailath, T.: Linear systems. Prentice-Hall, N.J (1980)
[41] Dorf, R., Bishop, R.: Modern control systems. Addison-Wesley, Reading (1995)
[42] Femat, R.: A control scheme for the motion of a magnet supported by type-II superconductor. Physica D 111, 347 (1998)
[43] Puebla, H., Alvarez-Ramírez, J., Cervantes, I.: A simple tracking control for Chuás circuit. IEEE Trans. Circ. and Syst. I 50, 280 (2003)
[44] Solis-Perales, G.: Sincronización de Marcha de Polípodos, McS. Thesis (in Spanish) (1999)
[45] Chou, C.C., Lauk, M., Collins, J.J.: The dynamics of quasi-static posture control. Human Movement Sci. 18, 725 (1999)
[46] Hall, K., Cristini, D.J., Tremblay, M., Collins, J.J., Glass, L., Billete, J.: Dynamic control of cardiac alternans. Phys. Rev. Lett. 78, 4518 (1997)
[47] Femat, R., Alvarez-Ramírez, J., Zarazua, M.A.: Chaotic behavior from human biological signal. Phys. Lett. A 214, 175 (1996)
[48] Pence, D.V., Beasley, D.E.: Chaos suppression in gas-solid fluidizatio. Chaos 8, 514 (1998)
[49] Teel, A., Praly, L.: Tools for semiglobal stabilization by partial state and output feedback. SIAM J. of Control Opt. 33, 424 (1991)
[50] Parmenada, P., Mena, C.H., Baier, G.: Resonant forcing of a silent Hodking-Huxley neuron. Phys. Rev. E 66, 047202-1 (2002)
[51] Esfandiari, F., Khalil, H.K.: Output feedback stabilization of fully linearizable systems. Int. J. of Control 56, 1007 (1992)
[52] Sussman, H.J., Kokotovic, P.V.: The peaking phenomenon and the global stabilization of nonlinear systems. IEEE Trans. on Automatic Control 36, 461 (1991)
[53] Chua, L.O., Yang, T., Zhong, G.Q., Wu, C.W.: Adaptive synchronization of Chua oscillators. Int. J. Bifur. and Chaos 6, 189 (1996)
[54] Torres, L.A.B., Aguirre, L.A.: Inductorless Chua's circuit. Electronics Letts. 36, 1915 (2000)
[55] Kothare, M.V., Campo, P.J., Morari, M., Nett, C.N.: A unified framework for study of anti-windupdesigns. Automatica 30, 1869 (1994)
[56] Rönbäck, S.: Linear control of systems with actuators constraints, Ph. D. Dissertation, Luleå University of Technology, Sweden (1993)
[57] Doyle III, F.J.: An anti-windup input-output linearization scheme for SISO systems. J. Proc. Control 9, 213 (1999)
[58] Alvarez-Ramírez, J., Garrido, R., Femat, R.: Control of systems with friction. Phys. Rev. E 51, 6235 (1995)
[59] Moon, F.C.: Chaotic vibration of a magnet near a superconductor. Phys. Lett. A 132, 249 (1988)
[60] Hikihara, T., Moon, F.C.: Chaotic levitated motion of a magnet supported by superconductor. Phys. Lett. A 191, 279 (1994)
[61] Chang, P.-Z., Moon, F.C., Hull, J.R., McCahly, T.M.: Levitation force and magnetic stiffness in bulk high-temperature superconductors. J. Appl. Phys. 67, 4358 (1990)
[62] Goodall, R.M., Maclod, C.J.: Proc. of the 4TH IEEE Conference on Control Appl., Albany N.Y, p. 261 (1995)
[63] Chen, Y.H., Chou, M.Y.: Continuous feedback approach for controlling chaos. Phys. Rev. E 50, 2331 (1994)

[64] Chen, G., Dong, X.: On feedback control of chaotic continuous-time systems. IEEE Trans. Circuits and Systems 40, 591 (1993)
[65] Aström, K.J., Witterman, B.: Adaptive Control. Addison-Wesley, N.Y (1989)
[66] Dorato, P.: Robust control. IEEE Press, N.Y (1987)
[67] Bryksin, V.V., Dorogovtsev, S.N.: Nonlinear diffussion of magnetic flux in type-II superconductors. JETP 77, 791 (1993)
[68] Moon, F.C., Wenf, K.-C., Chang, P.-Z.: Dynamic magnetic forces in superconducting ceramics. J. Appl. Phys. 66, 5643 (1989)
[69] Bowong, S., Moukam-Kakmeni, F.M.: Chaos control and duration time of a class of uncertain chaotic systems. Phys. Letts. A 316, 206 (2003)

3 Robust Synchronization of Chaotic Systems: A Proportional Integral Approach

In previous chapter the chaos suppression was discussed. However, there is one more interesting problem in chaos control: the synchronization. Synchronize means to share the same time and signifies that two or more events occurs at same time. In nonlinear science diverse synchronization phenomena have been found in chaotic systems. Thus, such a problem results in very interesting dynamical phenomena and has technological applications, as in communication [1], and scientific impact as, for example, in animal gait [2],[3] or cells of human organs [4]. A continuation path for synchronization is in spatially extended systems [5] where synchronization phenomena are already being studied. Other interesting issues on synchronization is, on the one hand, the cost of synchronizing chaotic systems [6]; that is, to measure the energy required to achieve chaotic synchronization. Here, the control theory can be exploited to include cost function at design of synchronization command by computing optimal, sub-optimal and/or robust controllers [7]. On the other, the geometrical properties of synchronization are also a raising theme [8], [9]. Here, geometrical control theory can be used to compute the invariant manifolds [10]. This Chapter is related to the robust synchronization, and is centred on the robust analysis and some interpretations about robustness in synchronization. To this end we exploit the simpler controller in Chapter 2: the Proportional-Integral feedback and some approaches.

3.1 Chaos Synchronization Via Linear Feedback

Roughly speaking, the chaos synchronization problem consists of making two or more chaotic systems oscillate, in some sense, in synchronous way. This could seem impossible, if we think on the following features about the chaotic motion: (i) The sensitive dependence on initial conditions. It is almost impossible to reproduce the same starting conditions on two (or more) chaotic systems. (ii) To match exactly the model parameter of two (or more) chaotic systems. Even infinitesimal variations of any model parameter will eventually result in divergence of chaotic orbits starting nearby each other. However, chaos synchronization has been reported by several authors.

R. Femat & G. Solis-Perales: Robust Syn. of Chaotic Sys. Via Feedback, LNCIS 378, pp. 51–97, 2008.
springerlink.com

Since chaos synchronization has potential applications in several areas, the problem has attracted attention of many researchers. Some examples of synchronization application are such as: secure communication [11], [12] or biological oscillations like the animal gaits [3], or neurons clusters [13]. Up today, the chaos synchronization has been mainly attained via the information feedback and feedforward. For instance, it has been demonstrated that two similar chaotic oscillators (same model, but different parameters), can be synchronized using high-gain feedback; see [14] and references therein. It should be pointed out that high-gain feedback can yield very sensitive feedback actions and small variations could induce an unstable behavior. Adaptive techniques have been reported to achieve the synchronization for similar chaotic oscillators [15], [16]. The synchronization can be achieved in spite of the parameter have random motion. However, it requires to know the explicit dependence (multiplicative or additive) of the system on parameter structure. With this kind of controllers, one can obtain a small magnitude error. The error has high frequency oscillations, which can affect the closed-loop performance (this is, since every variation in the error, it could be required a large amount of feedback energy to maintain the synchronization, for details see [17]). In the case of the information feedforward, two identical systems have been considered. One of them is the master and full model is considered while another is the slave, which is a reduced model of the master, in other words a subsystem of the master system is constructed. In order to synchronize them the transmitted signal is the state of the master corresponding to the removed state in the slave. This information fed-forward technique has been used in secure communication by means of parameter variation, in this case the information is modulated by a parameter [18].

Recently, it has been reported that two strictly different chaotic oscillators (different model and same order) can be synchronized by means of a robust asymptotic feedback (RAF) [19]. Under RAF, the synchronization error converges to a bounded region, which contains the origin. This strategy requires a minimum information about the model of the master system. Such a control scheme consists of two parts: the first one is a linearizing-like control law and the second one is an uncertainties estimator. In similar manner than PI and PII^2 controllers in previous chapter, the uncertainties are lumped in a nonlinear function such that the controller does not require the structure of the master system. The nonlinear lumping function can be interpreted as a state of a equivalent dynamically system. The estimation of the uncertainties is obtained via the augmented state, which results in the increment of the order of the controller. Such control strategy has an adaptive structure and can achieve the synchronization of high order oscillators with the same model and different parameters [20]. Nevertheless, this implies the knowledge of any priori information to construct the feedback; for instance, the order of the synchronization system. However, it is desired a feedback scheme that: (i) requires least prior knowledge about the synchronization system (ii) the controller results in a low-order feedback, and (iii) the synchronization can be attained in spite of parametric variation and noisy environment.

Here, the synchronization problem of second-order driven oscillators with strictly-different model is discussed. The synchronization is achieved by means of a feedback scheme in the Laplace domain. The underlying idea is to show similitude and differences between synchronization and suppression of chaos in the context of robust control approach to geometrical control. In the sense, as in previous chapter, this controller requires to know *neither the internal nature of the model nor the structure parameter*. In addition, regarding the results in previous chapters, the Laplace controller has the following advantages: a) it requires least prior information about the systems to be synchronized, and b) the order of the Laplace-domain controller is less than previously reported. In this way, this chapter promotes the following basic ideas: (i) chaos synchronization of second-order driven and third order oscillators can be achieved by linear feedback, (ii) Chaos synchronization can be attained in spite of least prior information about the master system, and (iii) Synchronization can be achieved even noisy measurements environment. As was presented in the Chapter 2 the Laplace controller can be constructed with three parts; a proportional action, an integral action and a quadratic integral action. All of them result in a second order feedback. The first one provides a signal, which is proportional to the synchronization error. The second one is a reset action. The combination of these parts are so-called Proportional-Integral controller (PI). The last one has been recently developed. This part is able to estimate unknown terms (like external disturbances, modelling errors, parametric variations, etc.), and take control actions, therefore, this controller is robust against this class of uncertainties [21]. Here, with the proposed controller, an estimated value of the synchronization error is computed, which is dynamically obtained for the Laplace controller. In this way, it is possible to achieve the synchronization of a chain of strictly different chaotic oscillators (different model) with the advantage that the controller does not need prior knowledge about the structure of the master (drive) model.

For the sake of completeness, it should be pointed out the importance of the synchronization of systems with strictly different model. Let us consider, for example, the communication between neurons and muscles. It is expected that the neuron and the muscle models be strictly different. However, the normal muscle motion is synchronous, why?. In our opinion simple feedback structures allow this synchronization and its study as feedback control can leads us to knowledge about this and other systems. Thus, this kind of approach can have potential application in biology, mechanical devices, etcetera. Moreover, for instance, Rabinovich and coworkers (see [13]) have studied the synchronization in different neurons of a class of crustacean. In this sense, the study of synchronization of strictly different oscillators can provide timely and promissory results. Here we show an alternative scheme to study the chaos synchronization of strictly different oscillators.

3.1.1 The Feedback Control Scheme

The synchronization can be addressed from two different points of view: the first one is a tracking problem, which consists in make the trajectories of the slave system follow the master trajectories, i.e. $x_{i,S} \rightarrow x_{i,M}$, $i = 1, 2$ (see for instance [8]). To this end, the master dynamics should be known; at least throughout an output signal. The second one can be addressed as a stabilization problem; i.e., the reference for the closed-loop is the zero point. Both formulations are similar in the sense that the control objective is to hold the discrepancy $\| x_{M,i} - x_{S,i} \|$ around zero for any initial condition and all time after control is turned on. Nevertheless, since in the latter (stabilization formulation) a dynamical system is constructed from the difference between the master minus slave systems [22], the new system is stabilized around origin such that $x_i \triangleq (x_{M,i} - x_{S,i}) \rightarrow 0$ for all $t > 0$, $i = 1, 2, 3, \ldots, m \leq n$ and $x(0)$ belonging in an open subset U of $\mathbb{R}^2$. In this book the synchronization problem is interpreted as chaos stabilization one; i.e., the system to be stabilized is constructed from the discrepancy between master and slave system, see Fig. 3.1. Here, we also show that it is possible to obtain simple models for feedback controllers in Laplace domain, which results in input-outputs models [21]. Such models provide information about the output response of the dynamical system in such a manner that optimal conditions can be imposed on performance or stability [23]. Here, in order to derive the stabilization, the transfer function of the chaotic synchronization system is not computed. Hence, although the controller in block C in Figure 3.1 is linear, the dynamical systems in block P remains nonlinear.

Figure 3.1 shows a typical feedback control system. The system output y is a measurable state, which represents a natural trajectory of the dynamical system into the block P. The objective of the block C (controller) is to steer the trajectories of the system P in such way that they track the reference signal, r. The internal structure of the block C can be linear or nonlinear. If the control command, u, is acting onto the system P, from the return of a portion of the information of the output y, then it is said that the feedback loop system is closed. The difference between the master and slave signals, $E = r - y$, is the unique signal that the controller uses to perform the stabilization. This is a practical advantage. In the classical control schemes, each block represents a transfer function (i.e., the input-output representation). Thus, if structure into C is linear, the controller transfer function is given by $C(s) = u(s)/E(s)$. However, note that to compute a transfer function for a chaotic (nonlinear) systems does not make sense. This is the case for nonlinear structures into C.

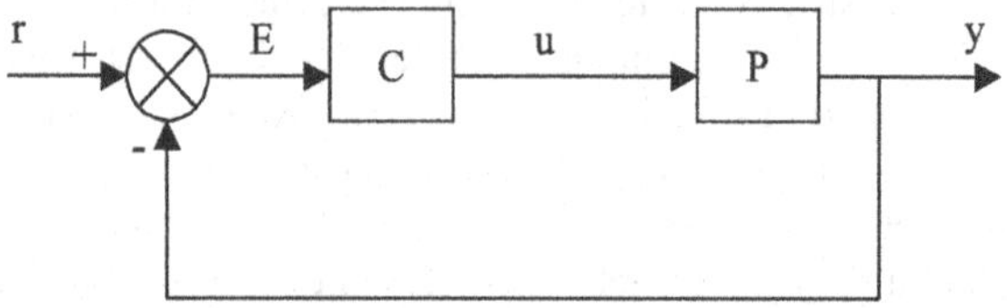

Fig. 3.1. Typical feedback control system. Where y is the system output, u is the control command, E is the error signal, r is the desired reference and C is the controller and P is the process.

Exercise 3.1. *Closed-loop dynamics of a linear system under PII^2 control.* Assume that the structure into block P (plant) is linear, and given by the transfer function $P(s) = K_p/(\tau_p s + 1)$, where K_p and τ_p are strictly positive constants. Moreover, suppose that the structure of C is given by Eq. (2.13). (i) Find the values of the control parameters such that stability can be assured, and (ii) Show the closed-loop dynamics in classical terms (i.e., from, for example, the Bode plot).

3.1.2 The Chaos Synchronization as a Stabilization Problem

Consider two second order dynamical systems with the following structure

$$\dot{x}_M = F_M(x_M; \pi_M) + T_M(t) \tag{3.1}$$

$$\dot{x}_S = F_S(x_{S;}\pi_S) + T_S(t) + Bu \tag{3.2}$$

where $x \in \mathbb{R}^2$, $F_i(x_i;\pi_i)$ are smooth nonlinear vector fields, $T_i(t)$ denote the external excitation force and $\pi_i \in \mathbb{R}^p$ are parameter sets of the system, for i = M,S. It is well established that, depending on the parameter values and the external excitation, systems (3.1) and (3.2) can display a chaotic behavior [24]. It is important to note that systems to be synchronized can have different, (i.e. $F_M \neq F_S$), or identical, (i.e. $F_M \equiv F_S$), models.

Now, the synchronization problem can be stated considering the following:

S.3.1) Only one state of each system is available for feedback.
S.3.2) The external perturbing forces $T_i(t)$, where i = M,S, are smooth and bounded.
S.3.3) The control channel corresponds to the measured state.

Assumption *S.3.1)* is not restrictive. Indeed, in most of nonlinear systems only one state (for instance, time series) is available for feedback. Moreover, it is desirable to get as less as possible measured states for physical implementation. Assumption *S.3.2)* is also physically practical because, without lost of generality, we can claim that continuously (or, at least, sufficiently) differentiable terms models the dynamics in nature and artifacts. Assumption *S.3.2)* signifies that the external disturbances do not have brusque changes along time. Assumption *S.3.3)* is the more restrictive because it implies that relative degree of the synchronization system is equal to one. The relative degree is the lowest integer number such that the time-derivative of the system output, y, is directly related with the control command, u (see [25] for definition and [10] for technical details and implications on differential geometry). Nevertheless, this assumption is not restrictive. In fact, it has been reported that, the chaos suppression problem can be addressed even if the system has relative greater than one [20]. To this end, the unmeasured states can be considered as internal disturbance [20] or internal dynamics [10], [25]. Physically, Assumption (S.3.3) means that the control command acts directly onto the measured state, which is practical for many devices (e.g., chaotic circuits).

Now, following the ideas reported in [20] and [26], the dynamics of synchronization error can be computed from the difference between systems (3.1) and (3.2). That is, by defining $x := x_M - x_S$, the dynamical system $\dot{x} = \Delta F(x_M, x_S;\pi) + \Delta T(t) - Bu$; can be derived, where $\Delta F(x_M,x_S;\pi) = F_M(x_M;\pi_M) - F_S(x_S;\pi_S)$ and $\Delta T(t) = T_M(t) - T_S(t)$, π is a parameter set which lumps the differences between parameters of the response and drive system. In this sense, the synchronization problem can be stated as the stabilization of the synchronization error system at origin. This suggests that synchronization of system (3.1) and (3.2) can be achieved via linear feedback. There are some published results where synchronization is achieved via linear feedback [2]. However, as we shall see in Sect. 3.4, linear feedback yields local stability which implies that only for close initial conditions of the drive and response systems the synchronization is achieved. Fortunately, since chaotic systems yield bounded trajectories, robustness properties can be extended for semi-global stability, i.e., for a sufficiently large subset of $\mathbb{R}^2$.

Exercise 3.2. *Dynamics of the discrepancy system.* Consider the Duffing and van der Pol equations under chaotic behaviour. By numerical simulations, show the dynamics of the discrepancy system. Note that for any initial condition in an open subset of the discrepancy plane, containing the origin, the trajectories holds bounded and smooth. Hint: Define the difference (x_M - x_S) in an open subset of $\mathbb{R}^2$ containing the origin. ∘

3.1.3 A PI Controller with Uncertainties Estimation

In order to design a PI feedback control, let us consider that synchronization error system can be written as the nonlinear system

$$\dot{y} = \Gamma(y) + f(y) + g(u + d) \tag{3.3}$$

where $y \in \mathbb{R}$ denotes the output of a given chaotic system (for instance, position of nonlinear oscillator or voltage in a chaotic circuit), $\Gamma(y)$ is a linear function whereas $f(y)$ is a nonlinear function, g is a non-zero constant and $d = d(t)$ is a time function which denotes a disturbance into the dynamical system (3.3). The control goal is to design a feedback control u, given by a PI transfer function, such that the scalar $y = y(t)$ holds close (tracks) a desired signal $r = r(t)$ for all time $t \in \mathbb{R}_+$. Notice that, in particular, since the chaotic synchronization is addressed as the stabilization at origin, the desired signal $r \equiv 0$ for all $t \geq 0$. Equation (3.3) implies that the output of a nonlinear system can be represented as a differential equation. Moreover, the form of (3.3) is not restrictive; indeed the most of the nonlinear systems can be rewritten in such a manner. Without of loss of generality one can assume that functions $d = d(t)$ and $f(y)$ are uncertain; then, in presence of additive uncertainties on parameters of linear terms, the equation (3.3) becomes $\dot{y} = \Gamma^*(y) + \Gamma^u(y) + f(y) + (g^* + g^u)(u+d)$, where $\Gamma^*(y)$ and g^* are known and given by nominal values while $\Gamma^u(y)$ and g^u are unknown. In this manner, equation (3.3) can be rewritten as follows

$$\dot{y} = \Gamma^*(y) + g^* u + \phi(y,d,u) \tag{3.4}$$

where $\varphi(y,d,u) := \Gamma^u(y) + f^u(y) + (g^* + g^u)d + g^u u$ is a unknown nonlinear function.

Here, the control goal for the synchronization systems is to design a PI-type feedback control u such that y tends to zero in spite of the model errors in the uncertain function $\varphi(y,d,u)$ (which lumps the unknown terms) and external disturbances $d = d(t)$. Then, the following second-order estimator can be used [26]

$$\begin{aligned}
\dot{\hat{y}} &= \Gamma^*(y) + g^* u + \hat{\eta} + \kappa_1(y - \hat{y}) \\
\dot{\hat{\eta}} &= \kappa_2(y - \hat{y}) \\
u &= -\frac{1}{g}\left(\Gamma^*(y) - \hat{\eta} + k_c y\right)
\end{aligned} \tag{3.5}$$

where $(\hat{y}, \hat{\eta})$ are estimated values of y and $\varphi(y,d,u)$, respectively, k_c is a positive constant which denotes the controller gain. Note that the compensator (3.5) is linear, hence, from Laplace operator, the transfer function of the controller, $C(s) = u(s)/E(s) = u(s)/(y(s) - r(s)) = u(s)/y(s)$, the control command u, after algebraic manipulations, becomes

$$C(s) = K_c\left(1+\frac{1}{\tau_I s}+\frac{k_e}{s(\tau_{II}s+1)}\right) \tag{3.6}$$

where k_c and k_e are the proportional gain and estimation gain respectively and this factors are considered as an amplification factor and provides stability to the system. τ_I, τ_{II} are time constants for the integration and the dynamic estimation respectively, the first one is considered as the integration time, and the last one is the estimation time. This controller is formed by the following parts:

(i) Proportional action. The acting output of this part is proportional to the error E = r - y, where y is the system output (measurement) and r is the reference (desired signal). Then, the proportional control action is given by $u \equiv u_s + k_c E$, where u_s is the bias signal of the controller (i.e., the acting output of the controller when E = 0; this is called the steady-state control command).

(ii) Integral action. The control command includes a reset time (the reset time is so-called integral time constant). The reset time is an adjustable parameter and is sometimes referred to repeat time. This control action makes the control command varies as long as the error signal changes. Therefore this controller can eliminate the offset if the reference is constant. Is very common to combine the proportional action and the integral action to get the classical PI controller.

(iii) The quadratic integral action. This term increases the ability of the dynamical estimation of the disturbances and uncertainties of the PI controller. The main objective of this action is to detect uncertain disturbances via dynamic estimation. In this way, the feedback (3.6) yields the tracking of a desired smooth signal.

Exercise 3.3. *Derive the transfer function of the PII^2 controller (3.6) departing from (3.5).*

The controller described by (3.6) is able to estimate dynamical uncertainties into the system to be controlled and external perturbations. This controller has low order and is linear. In addition, it does not require the structure of the system to be controlled because it is constructed from the time derivative of the output; it only requires the differences between the reference signal, r, and the measured state, y. On the other hand, the synchronization of strictly different oscillators can be achieved by means of an observer based controller [19]. However, such controllers can result in high-order feedback. For instance, in order to synchronize second-order driven systems the resulting controller results in third order feedback.

From Eq. (3.6), we can find that the control command has the following form

$$u(s) = k_C\left(1+\frac{1}{\tau_I s}+\frac{k_e}{s(\tau_{II}s+1)}\right)(y(s)-r(s)) \tag{3.7}$$

k_c is the proportional gain, k_e is the quadratic integral gain, and the constant τ_I and τ_{II} are the time constant of the controller; τ_I represents the reset time whereas τ_{II} is the characteristic time of the dynamic estimation. u(s) and E(s) are deviation variables (u(s) = u - u_s and E(s) = y(s) - r(s). k_c, k_e, τ_I and τ_{II} are selected in such way that the closed-loop system be stable (this is, the synchronization system under control actions).

In the context of the stabilization of the synchronization error, the output of the system, y, is given by the difference x_M - x_S, where x_M and x_S are the measured states of the driven and response system, respectively, i.e., $y = x_M - x_S$. In this way, the feedback control command is a coupling force, which leads the slave trajectories to the master trajectories. Thus the synchronization of two nonlinear driven oscillators results in the following coupled system

$$\begin{aligned} \dot{x}_{1,S} &= x_{2,S} \\ \dot{x}_{2,S} &= f_S(x_S;\pi_S) + \tau_S(t) + u \\ \dot{x}_{1,M} &= x_{2,M} \\ \dot{x}_{2,M} &= f_M(x_M;\pi_M) + \tau_M(t) \end{aligned} \tag{3.8}$$

the input u is given by Eq. (3.7), f_M, $f_S \in \mathbb{R}^2$, and as the tracking problem was interpreted as a stabilization problem then $r = 0$ and $y = x_{2,M} - x_{2,S}$. Thus, we have structured a new system, which addresses the synchronization problem. It should be pointed out that the system (3.8) does not depend on any specific $f_M(x_i;\pi_M)$ nor $f_S(x_i;\pi_S)$. In this sense, synchronization of two strictly different oscillators can be achieved.

3.1.4 Local Stability Analysis for an Illustrative Example

In seek of clarity, let us consider the van der Pol and Duffing model for stability analysis. To determine the control parameters such that the closed-loop system be stable, it is required to solve the following question: Is there any control parameter value such that the system (3.8) be, at least, locally stable?. From the following fact, local stability analysis is sufficient to argue semi-global stability.

Lemma 3.4. [24] Suppose that there exist external perturbing forces ($\tau_M(t)$ and $\tau_S(t)$) acting over the system (3.8), which are bounded and smooth. Then, there exist undecomposable closed invariant sets, Δ_M and Δ_S, with the following uncontrolled property: Given $\epsilon_M > 0$ and $\epsilon_S > 0$, there are sets U_M and U_S of positive Lebesgue measure in the ϵ_j - neighborhood ($j = M,S$) of Δ_j such that ω_j - limit sets given by $x_j(t)$ are contained in Δ_j and the forward orbit of $x_j(t)$ are contained in U_j, respectively.

Some comments about above result are in order. (a) A second-order oscillators cannot yield chaotic behavior. In fact, Poincare-Bendixon theorem [24] implies that two trajectories cannot cross in a two-dimensional state space. Hence, second-order systems should be forced. Such phenomena is reproduced in many mechanical, physical and biological systems. For example, a forced magnetic beam results in Duffing equation, which models a violin string. (b) Many physical, chemical, or biological forces are smooth and time-continuous. Thus, continuity and smoothness are not restrictive conditions. (c) A characteristic of the behaviour of chaotic systems is that it is formed by multiple unstable orbits which are contained into an invariant set, i.e., the time evolution of the system flows are bounded. For instance, consider the forced magnetic beam modelled by the Duffing equation, the beam oscillations are bounded.

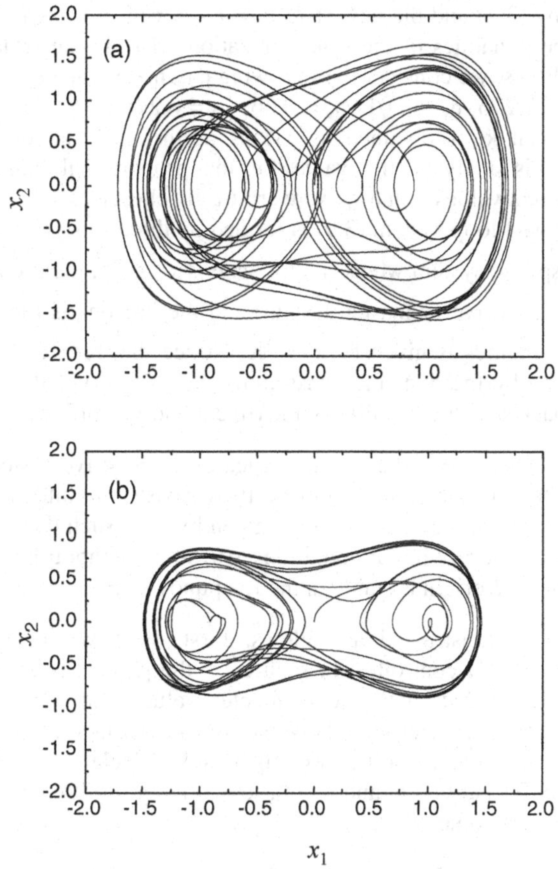

Fig. 3.2. (a) Phase portrait of drive system (van der Pol). (b) Phase portrait of the response system when u = 0 (Duffing).

Thus, solution of the chaos synchronization problem implies that the orbits (Fig. 3.2b) of the response system (van der Pol equation) track the orbits (Fig. 3.2a) of the drive system (Duffing equation). This means that the slave system must have the same behavior than the master system. The control command can be interpreted as a coupling force between the drive and response systems (note that, in this sense, the proposed controller is agree with synergetic theory [27]). Therefore, the local stability of the closed-loop system can be established as follows

$$\begin{aligned}
\dot{x}_{1,S} &= x_{2,S} \\
\dot{x}_{2,S} &= \gamma_D \cos(\omega_D t) + x_{1,S} - x_{1,S}^3 - \delta_D x_{2,S} + u \\
\dot{x}_{1,M} &= x_{2,M} \\
\dot{x}_{2,M} &= \gamma_V \cos(\omega_V t) - (x_{1,M} x_{2,M} - x_{2,M})\delta_V - x_{1,M}^3 \\
\dot{x}_3 &= x_{2,M} - x_{2,S}
\end{aligned} \tag{3.9}$$

where u is given by (3.7) and the new state $\dot{x}_3 = r - y$, with $y = x_{2,M} - x_{2,M}$ and $r \equiv 0$, which provides the dynamics of the synchronization error. In particular, for second-order driven oscillators, we consider the following parameter values of the master and slave system: ($\gamma_V = 1.275$, $\delta_V = 0.01$, $\omega_V = 0.36$; $\gamma_D = 0.275$, $\delta_D = 0.15$, $\omega_D = 1.1$. The value of the constant k_c is obtained by means of the analysis for a PI feedback controller. From this analysis, the value for the proportional gain, $k_c = 5.0$, is obtained. The constant values τ_I and τ_{II} were arbitrarily chosen as 1.

Note that the closed-loop system (3.9) is formed by three subsystems. The first one is given by the response system whose vector state is $x_S \in \mathbb{R}^2$, the second one has the vector state $x_M \in \mathbb{R}^2$ (drive or master system) and the last one is represented by the feedback whose dynamics is given by $x_3 \in \mathbb{R}$. Notice that both drive and response systems are coupled by the dynamic constraint, which is provided by the feedback. Besides, the feedback structure is very simple (linear and low order).

Proposition 3.5. Consider that the Duffing equation is the drive system and van der Pol oscillator is the response system then there exists an interval for dynamic estimation parameter k_e of the coupling force given by (3.7) such that the system (3.9) is locally stable; i.e., for any initial conditions $x(0)$ in a neighbourhood of the origin point the trajectories the origin is stable in the Lyapunov sense.

Proof. The prove is constructed as follows. First we have considered that the feedback is given by a PI controller. In such case, the parameter to be found is the control gain, k_c. After that, with the parameter values and according to the PI controller, we only have to determine the value for the constant k_e (quadratic integral gain). Therefore, we have to solve two eigenvalue problems. Each one for the Jacobian of system (3.9) at $t \to \infty$ and $t \to 0$, respectively. By taking the $\lim_{t\to 0} J$, the Jacobian of Eq. (3.9) becomes

$$J_1 = \begin{pmatrix} 0 & 1 & 0 & 0 & 0 \\ 1 & -\delta_D - (k_C - k_e\tau_{II}^2) & \left(k_c/\tau_I + k_c k_e\right) & 0 & (k_C - k_C k_e \tau_{II}^2) \\ 0 & -1 & 0 & 0 & 1 \\ 0 & 0 & 0 & 0 & 1 \\ 0 & 0 & 0 & 0 & -\delta_V \end{pmatrix} \tag{3.10}$$

where J_I represents the local dynamics of the system (3.9) at high-frequencies. From the matrix (3.10), we have to make the spectrum of J_I negative defined by choosing the values, k_{eI}. Indeed, we found an interval for the dynamic estimation parameter, k_{eI}, such that the system (3.9) is locally stable at high-frequencies. Now, taking the $\lim_{t\to\infty} J$, the following Jacobian can be obtained

$$J_2 = \begin{pmatrix} 0 & 1 & 0 & 0 & 0 \\ 1 & -\delta_D - k_C & \left(k_c/\tau_I + k_c k_e\right) & 0 & (k_C - k_C k_e \tau_{II}^2) \\ 0 & -1 & 0 & 0 & 1 \\ 0 & 0 & 0 & 0 & 1 \\ 0 & 0 & 0 & 0 & -\delta_V \end{pmatrix} \tag{3.11}$$

where J_2 represents the dynamics of the system (3.9) at low-frequencies. Once again, the parameter values, k_{e2}, are computed such that the matrix (3.11) has its eigenvalues at the open left-hand complex plane. This means that there exists an interval of the dynamic estimation parameter, k_{e2}, such that the system (3.9) is locally stable at low-frequencies. Finally, by continuity, the intervals can be intersected and one can obtain a new interval where the system (3.9) is locally stable for all $0 < t < \infty$. □

Remark 3.6. For the case of van der Pol (drive system) and Duffing (response system), the close interval, $k_{e1} \in [-0.8, 1.03]$, can be obtained for J_1 whereas for J_2, one can compute that $k_{e2} \geq -0.8$. Finally, taking the intersection, for any $k_e \in [-0.8, 1.03]$ the synchronization between van der Pol and Duffing oscillator is achieved and the closed-loop system is locally stable. The generalization of the result in Proposition 3.1 is given in last section of this chapter. Figure 3.3 shows the synchronization of van der Pol and Duffing oscillators, note that the stability holds against noisy environment. This fact is due to the filtering feature of the Luenberger observer. In this case, a noisy signal was added to the measured state, this is $y = x_{1,M} - (x_{1,S} + d)$. The noise considered was a random signal of amplitude 0.2. In this sense, formal study of robustness is currently an open problem. The H_2/H_∞ control theory can be exploited to design controllers towards robust stability and performance. This theory often yields

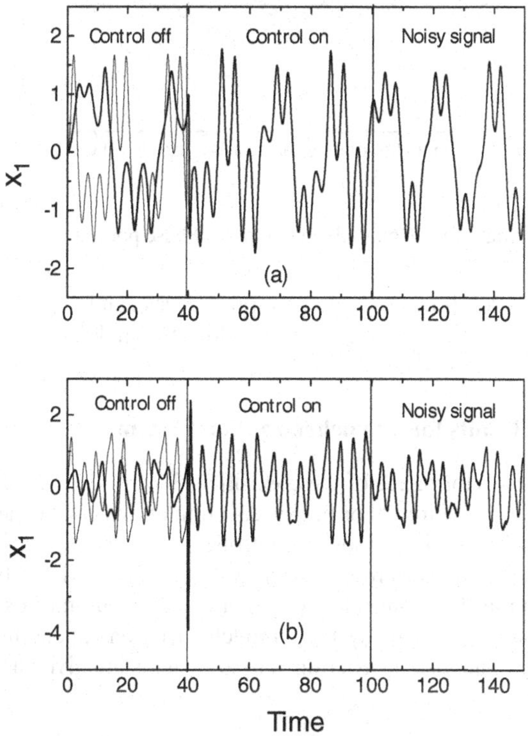

Fig. 3.3. Complete practical synchronization between van der Pol oscillator and Duffing oscillator, (a) synchronization between $x_{1,M}$ and $x_{1,S}$, (b) synchronization between $x_{2,M}$ and $x_{2,S}$

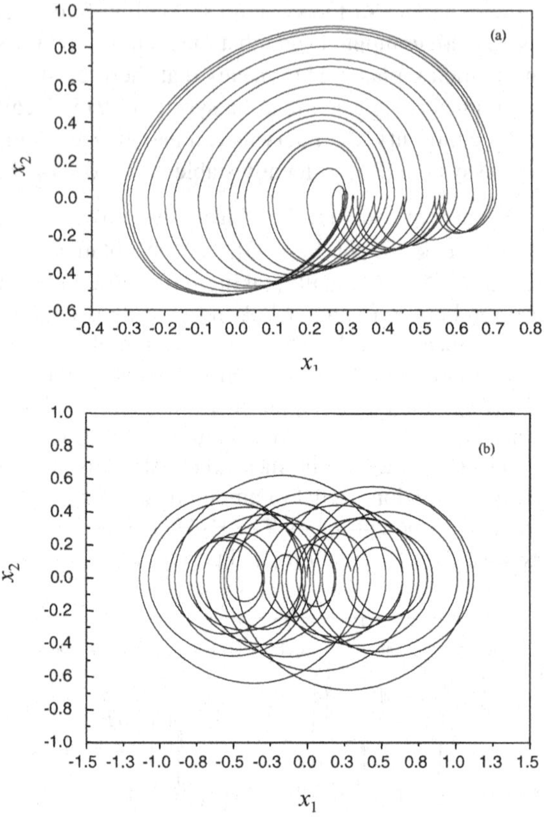

Fig. 3.4. (a) Phase portrait of the Coulomb equation, (b) Phase portrait or the pendulum equation

controllers of high order, however reduction procedures can be used to derive Laplace domain controllers (see [28] for the case of reduced complexity controllers in active suspension).

3.1.5 Numerical Results for a Synchronization System

In order to show the performance of the controller (3.7), or equivalently (3.5), chaotic systems can be realized in circuits or numerical simulations can be carried out as well. The aim is to show that the proposed controller can synchronize strictly different chaotic oscillators for second-order driven systems. Other illustrative examples of chaos synchronization by controller (3.7), are the Coulomb oscillator and the pendulum equation. Coulomb system models friction phenomena as stribeck phenomenon and dry and viscous friction. These systems are given as follows

$$\dot{x}_1 = x_2 \tag{3.12}$$

$$\dot{x}_2 = \gamma_{c} \sin(w_c t) - \alpha x_1 - \phi(x_1)\mathrm{Sign}(x_2)$$

$$\dot{x}_2 = -\delta_P x_2 - \sin(x_1) + \gamma_P \cos(\omega_P t) \tag{3.13}$$

where $\gamma_C = 1.77$, $\omega_C = 1.39$, $\alpha = 0.83$, $\varphi(x_1) = 1+kx_1$ for $x_1 > -1/k$ and $\varphi(x_1) = 0$ for $x_1 \leq -1/k$, and $k = 1.5$ for the Coulomb system (3.12), on the other hand, for the pendulum equation (see 3.13) $\delta_p = -0.13$, $\gamma_p = 0.5$ and $\omega_p = 1.0$. Figure 3.4a and Figure 3.4b show the phase portrait for these systems in open loop; i.e., without control.

To synchronize systems (3.12) and (3.13) as slave systems and van der Pol system as master one, first the appropriate control parameters must be found. To this end, a stability analysis is performed and the control parameters are as follows, for the Coulomb equation $k_c = 5$, $k_e = 1$ and for pendulum $k_c = 9$, $k_e = 5$. Figure 3.5 shows the phase portrait of both slave systems and that of the van der Pol after the control (3.5) is turned on. Notice that in both cases the phase portrait of Coulomb and pendulum equations have practically the same behaviour than the van der Pol equation, in other words they are practically synchronized [29]. This can be achieved because the feedback controller (3.6) absorbs (or provides) the required energy, i.e., the controller (3.5) excites some frequencies in slave system, in order to it behaves in synchronous way (see Chapter 2).

Figure 3.6a shows the synchronization between $x_{1,M}$ and $x_{1,S}$ here the control action was turned on at t = 40 s. Complete practical synchronization is achieved by the controller (3.6), dashed line represents slave system. In Figure 3.6b the synchronization

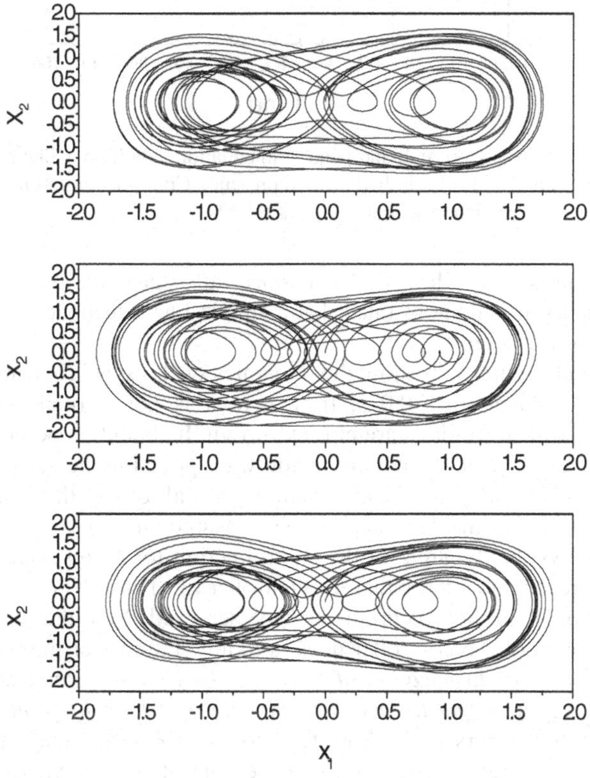

Fig. 3.5. Phase portrait of synchronized systems a) van der Pol, b) Coulomb, and c) Pendulum

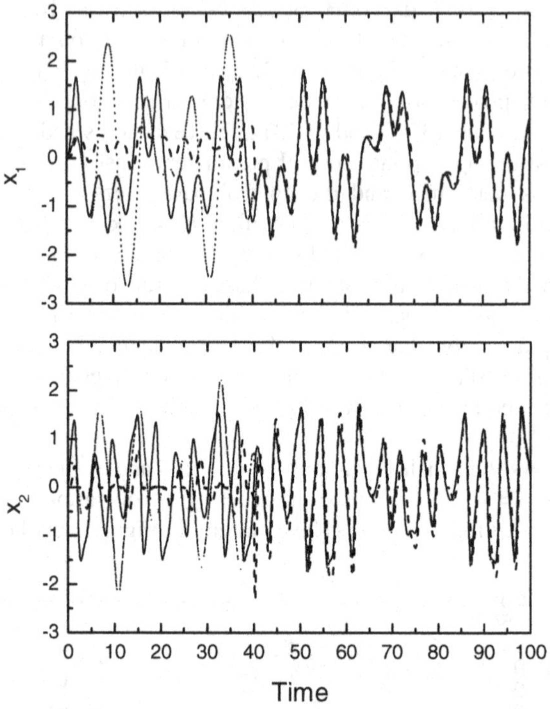

Fig. 3.6. Synchronization of Coulomb and pendulum equation, continuous line represents master system evolution (van der Pol), dashed line represents Coulomb equation and doted line pendulum equation, Control action was activated at t = 40sec.

between $x_{2,M}$ and $x_{2,S}$ is shown. Then since all states are complete practical synchronized, hence it is demonstrated that a simple linear feedback can synchronize second order chaotic systems.

Figure 3.7a and Figure 3.7b show the dynamics of the synchronization error. The control was activated at t = 40.0 s; from here it is not difficult to see that the synchronization error trajectories are close to origin. It should be pointed out that the measured state was x_2 while the control action was applied in the same state, x_2. This implies that the relative degree of the system is ρ equal to one (the relative degree is equal to the order of the time derivative of the measured output).

Figure 3.8 shows the dynamic of the control action for corresponding oscillators. Such signals represent the energy that the controller (3.5) absorbs (or provides) to the slave system in order to the synchronization be achieved. Note that, synchronization was attained in spite of the controller does not know the master model. It should be pointed out that *the synchronization of second order driven systems is achieved via controller (3.6) even if the master and slave model are strictly different*. Nevertheless, both synchronization errors, x_1 and x_2, tend to an arbitrarily small neighbourhood around the origin. Therefore, it is said that the controller (3.5) yields the complete practical synchronization via the control command [29].

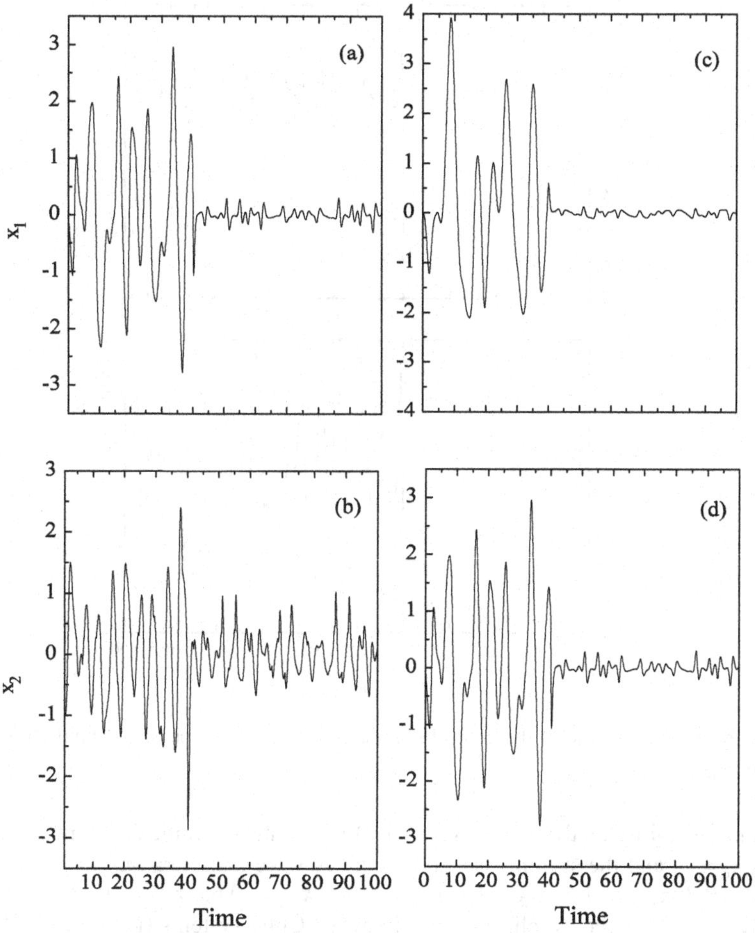

Fig. 3.7. Dynamic evolution of the error between van der Pol equation and Coulomb (a,b), and pendulum equation (c,d)

To illustrate the effect of time varying parameter in the synchronization of high-order systems via the controller (3.6); an illustrative case was considered. For this, the Chua's oscillator with different parameters is considered

$$\begin{aligned}\dot{x}_{1,i} &= \pi_{1,i}(x_{2,i} - x_{1,i} - f(x_{1,i}))\\ \dot{x}_{2,i} &= x_{1,i} - x_{2,i} + x_{3,i}\\ \dot{x}_{3,i} &= -\pi_{2,i}x_{2,i}\end{aligned} \tag{3.14}$$

where $f(x_{1,i}) = \pi_{3,i}x_{1,i} + ½(\pi_{4,i} - \pi_{3,i})(|x_{1,i} + 1| - |x_{1,i} - 1|)$ and i =M, S. Now, considering the Assumptions stated in Section 3.1.2 and the measured state is x_1 and the control action (3.7) is applied in the same state, so that the relative degree is

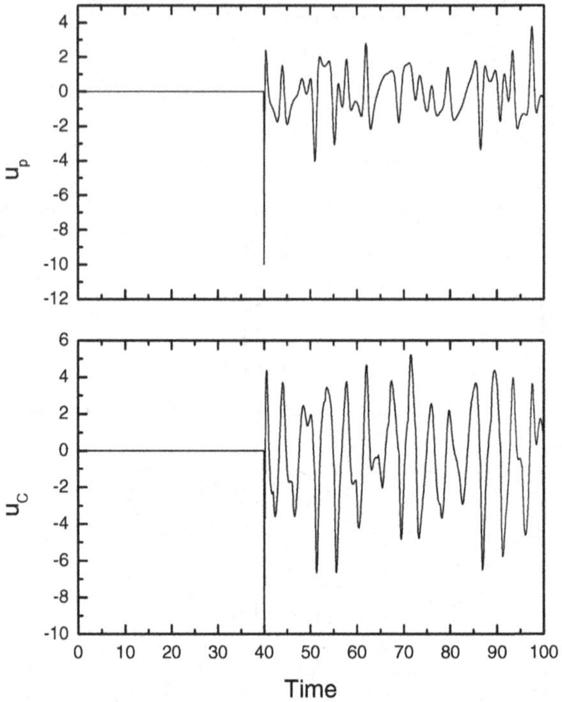

Fig. 3.8. Dynamic evolution of the control command for the Coulomb and Pendulum equation respectively

one. From a stability analysis (see Sec. 3.1.4) the control parameters were chosen as: $k_c = 10$, $k_e = 5$; and the systems parameter values were chosen as: $\pi_{1,M} = 10.2$, $\pi_{2,M} = 15$, $\pi_{3,M} = -0.8$, $\pi_{4,M} = -1.5$; $\pi_{1,S} = 10$, $\pi_{2,S} = 14.87$, $\pi_{3,S} = -0.68$, $\pi_{4,S} = -1.27$.

Figure 3.9 shows the synchronization between Chua systems (i.e., two third order chaotic systems) carried out by the controller (3.5). The control action was activated at t = 50 s. This illustrates that the feedback (3.5) is also able to synchronize high-order systems with different parameters.

3.1.6 The Performance Index and Tuning Procedure

From our proposal, the main objective of the feedback control is to lead the synchronization error dynamics around zero. As we have shown, this goal can be achieved by controller (3.5); or equivalently (3.7). Now, a measure of the control effort is desirable. The point is, the synchronization of two second-order driven chaotic oscillators is achieved via the controller (3.5), in spite of the master and slave system have strictly different model, against parametric differences and/or in face to noisy measurements. But, can the controller be tuned in such way that the control signal be, in some sense, small?. This is, to find the suitable parameters values, K_c and K_e, in order to the norm of the norm of the control signal satisfies performance requirements. To this end, a performance index, I_P, can be defined as follows

$$I_P = \int_0^t \left| x_{2,M} - x_{2,S} \right|^2 d\tau + \int_0^t \left| u_s - u \right|^2 d\tau \tag{3.15}$$

where u_s is a nominal control command (which corresponds to the steady state of the synchronization error). It is important to remark that the performance index (3.13) is related to the H_2-norm. The H_2-norm describes the average energetic accumulation of a signal [17]. The first integral in eq. (3.13) provides a measure of the synchronization error whereas the second term is the energy required to lead the trajectories of the synchronization error to origin. The performance index (3.13) can be interpreted as a measure of the synchronization error, which is weighted by the feedback energy. We also have to say that different index performance can be defined [30]. Once again, let us consider the van der Pol oscillator as the master system. Figure 3.10 shows the error norm and the control action norm for different values of k_c and k_e when the slave system is the Duffing oscillator. One can observe in Figure 3.10a that, as the quadratic action gain k_e, increases as the error tends to zero. However, the magnitude of the control action increases. Figure 3.10c shows the synchronization error norm. Note

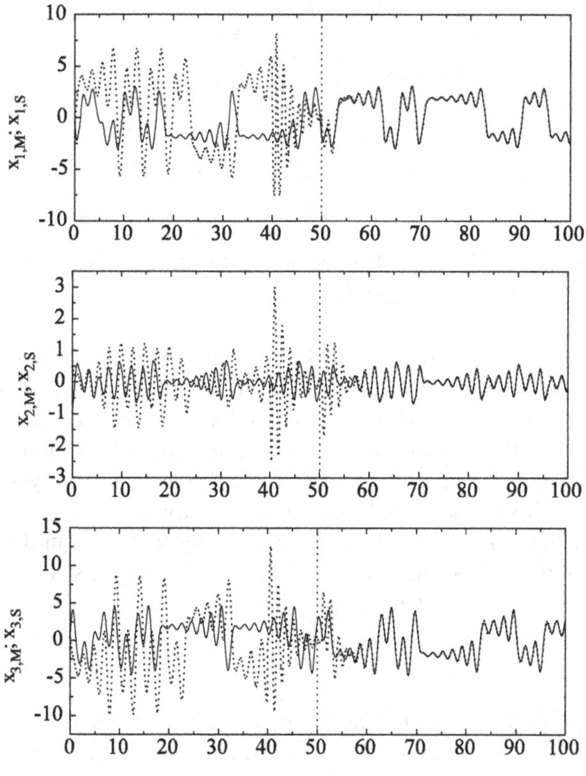

Fig. 3.9. Synchronization between two third order chaotic oscillators, control was turned on at t = 50s

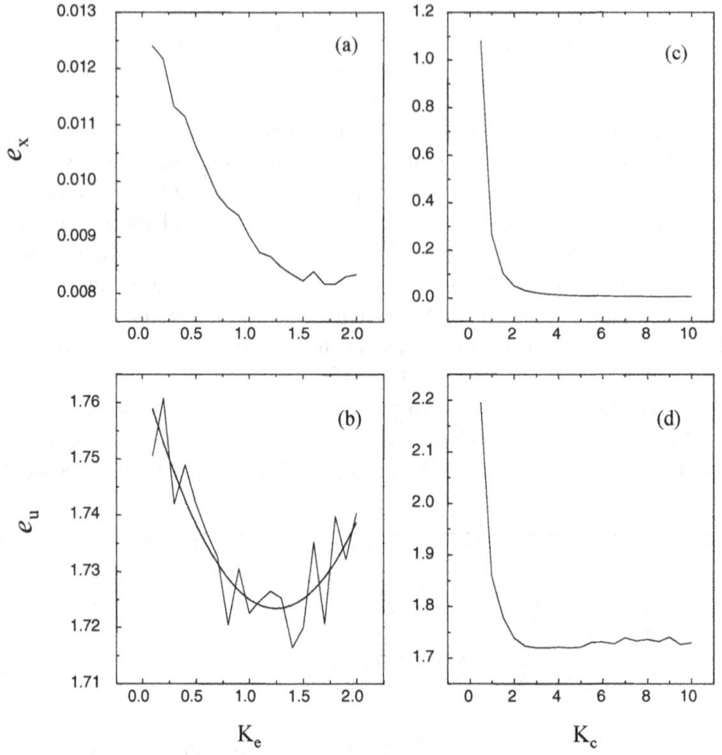

Fig. 3.10. Performance of the controller for the synchronization between van der Pol and Duffing oscillators, with $e_x = x_{2,M} - x_{2,S}$ and $e_u = u_S - u$

that it tends to zero as the proportional-integral action gain increases. Besides, the control action norm has little increases when the value for k_c is greater than 5 (see Figure 3.10d). This implies that the value of the proportional gain, k_c, does not increase considerably the feedback effort. Indeed, the control action does not increase as k_c increases. A disadvantage is that the controller becomes very sensitive which could induce instability to the system (see stability analysis section).

Now, to show the tuning procedure, let us consider that the van der Pol oscillator as the master system whereas the Duffing oscillator is the slave system (see eq. (3.9)). Assume that the controller is given by the following equation

$$C(s) = k_c(1+\frac{1}{\tau_I s}) \tag{3.16}$$

where k_c is the proportional control action gain and τ_I is a reset parameter. Note that the controller (3.14) does not comprise the dynamic estimation. In fact, it contains only two parts: (i) Proportional action and (ii) Integral action. Now, since C(s) = u(s)/E(s), the control action is given by

$$u = k_c E + \frac{k_c}{\tau_I} E \tag{3.17}$$

where $E = x_{2,M} - x_{2,S}$. From Proposition 3.1, there exists an interval for the parameter k_c where the system (3.9) is locally stable. Then, combining system (3.9) and Eq. (3.15), one can obtain the following Jacobian matrix

$$J = \begin{bmatrix} 0 & 1 & 0 & 0 & 0 \\ -1 & -(\delta_D + k_c) & -k_c & 0 & k_c \\ 0 & -1 & 0 & 0 & 1 \\ 0 & 0 & 0 & 0 & 1 \\ 0 & 0 & 0 & 0 & -\delta_v \end{bmatrix} \tag{3.18}$$

From the matrix (3.16), one can compute an interval for K_c in such way that the spectrum of J be negative defined. Finally, choosing the values of the damping parameters, δ_D and δ_V, in such way that the master and slave system display chaos. Hence, from the characteristic equation of the matrix (3.16) one has that $k_c > 1$. Note that the local stability only depends on the damping coefficients of both master and slave oscillators.

3.2 Chaos Synchronization Phenomena: A Classification

Two basic problems of the chaos control can be identified. These basic problems are the following: (i) chaos suppression and (ii) chaos synchronization. As we said above, chaos suppression mainly consists *in the stabilization of a system around regular orbits or equilibrium points.* The chaos synchronization problem has the following feature: *The trajectories of a slave system must tracks the trajectories of the master system in spite of both master and slave systems be different.* There are some uncertainties sources in the chaos control problem. Here we are interested to focus the uncertainties source of the synchronization problem. In this case the synchronization uncertainties comprises:(i) high-sensitivity of initial conditions, (ii) model mismatches and (iii) time-varying parameters. In regard of the above uncertainties one can observe the following:

(i) High-sensitivity of the initial condition (HSIC) [31]. Chaos definition is closely related to the concept of lost of the correlation between past and future, this means that small errors in any initial condition will extrapolate to large errors in trajectories. This concept is also related, in some sense, with chaos definition based on entropy or Lyapunov exponents [32]. In the context of the chaos synchronization, HSIC implies that discrepancy between master and slave systems at starting condition will result in a nonzero discrepancy along time. In addition, lost of the past/future correlation implies unpredictability and unreconstructibility. That is, it is almost impossible to predict the trajectories from an initial condition and viceversa, initial conditions cannot be constructed backward trajectories.

(ii) Model mismatches: There are two kinds of model mismatches (a) parametric and (b) structural. Parameter mismatches arises from different values of parameters of the slave model respect to master system. This kind of uncertainties is important for any synchronization application (for instance, secure communication).

Structural mismatches are related with the synchronization of strictly different oscillators [33] [39].

(iii) Time varying parameters: Even the parameter of the slave and master systems have the same nominal value, the time variation can be different. This is due to impairment on physical devices (for instance, capacitances in the Chua circuit, or damping factors in driven oscillators).

In spite of presence of uncertainties, the synchronization problem has been addressed with promissory results. Therefore, robust control schemes are required to achieve synchronization (for instance, see [34],[19].) The chaos synchronization phenomenon was found in earlier 90's. Pecora and Carrol [35], presented the first chaos synchronization results regarding the unidirectional synchronization of two identical chaotic systems (*i.e.*, two dynamical systems whose parameters are equal). They reported that as the differences between system increases, the synchronization is lost. Nevertheless, as the synchronization phenomenon has been understood, the researchers have found that synchronization of non identical systems can be achieved (see for instance [19] and references therein). From the paper by Pecora and Carroll [35], many papers have been published. Since the synchronization phenomenon is very interesting and important, a lot of effort have been devoted to understand it. For example, Phase Synchronization in bidirectionally coupled systems was reported in [36]. This is, two nonidentical systems have the same phase but different amplitude. We have found that phase synchronization can be attained in several synchronization phenomena. For instance, the practical synchronization (see below) is a class of synchronization where the systems have same phase but different amplitude. Also, in the partial synchronization [37], the systems can have the same phase and different amplitude. Nevertheless, the relationship between the different kinds of chaos synchronization (and the scenarios between them) have not been addressed yet.

This section is focused on the classification of chaotic synchronization when it is achieved via linear feedback. The main goal is to give a classification of the phenomena involved with the chaos synchronization problem. The results are restricted to the case where the systems to be synchronized are unidirectionally coupled by feedback. However, a general features are discussed. The discussed phenomena are: Complete Synchronization (CS), Partial-state Synchronization (PS), Practical Synchronization (PrS), Exact Synchronization (ES), Almost Synchronization (AS), Frequency synchronization (FS), and Phase Synchronization (PhS) [29][38].

In order to illustrate the synchronization phenomena a feedback control scheme is chosen. The controller has a simple (linear) structure. Indeed, it is represented by a Laplace domain equation, which implies that the feedback is linear. The controller contains three parts: (i) A term proportional to the control error, (ii) an integral action which provides stability around an equilibrium point and (iii) a quadratic integral action which yields a dynamic estimated value of the perturbing forces (for example, time-varying reference, internal perturbations or parametric variations). The scheme is so-called PII^2 controller (Proportional-Integral-Quadratic-integral). It has been shown that the chaos control problem can be addressed by means of the PII^2 controller [21]. In some sense, the synchronization of chaotic systems via feedback is related to bifurcation control [39]. It has been shown that bifurcation of a feedback controlled system yields several behavior [40][41]. Since synchronization of chaotic systems can be addressed from stabilization of the synchronization error, one can expect that synchronous behavior can be affected by bifurcation on the controlled systems.

Since the controller has a simple structure, hence, it is very easy to analyse its effect onto the chaotic system. The results allow to observe the different synchronization phenomena. In addition, the results show that it is possible to find a combination of phenomena in synchronized systems.

3.2.1 Some Synchronization Phenomena

In order to discuss different synchronization phenomena, let us consider the synchronization system, which is composed by two subsystems as in eq. (3.8). Roughly speaking, synchronization means the time-correlated behavior between both subsystems. In this sense, synchronization results from comparison of the properties of driven and response systems. To this end, it is said that the time invariant function h: $\mathbb{R}^m \times \mathbb{R}^m \rightarrow \mathbb{R}^m$ compares the measure behavior of both driven and response systems. Then we have the following definition

Definition 3.7. It is said that two chaotic systems are synchronized on their trajectories if there is a time-invariant mapping h: $\mathbb{R}^m \times \mathbb{R}^m \rightarrow \mathbb{R}^m$ such that $\| h(x_M, x_S) \| \leq \gamma < \infty$, where $\| \cdot \|$ represents any norm of Euclidean spaces, and x_M, x_S are the states of the drive and response systems.

Remark 3.8. The above definition is a general description of the synchronization phenomenon. Thus, according to Definition 3.7, the drive and response systems are synchronized if $\| h(x_M, x_S) \| = 0$ on all their trajectories. However, as we will see below, the synchronization phenomenon is quite complex. Indeed, we will show that even if there exist a function h: $\mathbb{R}^m \times \mathbb{R}^m \rightarrow \mathbb{R}^m$, the drive and response systems are not necessarily completely synchronized.

Chaotic synchronization has been classified as generalized, lag, or phase synchronization. A lot of efforts have been devoted to classification of the chaotic synchronization. Similar cases can be found from all of these classifications. Each synchronization phenomenon can be separately found into the same dynamical system, *i.e.*, since each synchronization phenomenon has different nature; however, it can be displayed by the same dynamical system. In the next paragraphs, we show that a combination of the synchronization types can be presented by the same system.

Exact Synchronization

Definition 3.9. It is said that two chaotic systems are exactly synchronized if the synchronization error, $x_i = x_{i,M} - x_{i,S}$, exponentially converges to the origin. This implies that at a finite time $x_{i,S} = x_{i,M}$, for i=1,2,..,m, *i.e.*, $\| h(x_M, x_S) \| = 0 \ \forall \ \tau > t_o$ and $x_M, x_S \in \mathbb{R}^m$.

Remark 3.10. Several controllers have been reported in the literature under perfect assumption. This is, the synchronization model is exactly known (see for example [42] and references in [43]). The above definition is general. It involves open-loop as well as closed-loop control schemes. Nevertheless, by invoking the Definition 3.7, it has been remarked by Aguirre and Billings [43] that, under feedback control actions, the exact controller has the following practical difficulties: (i) The model describing the system dynamics should be available in order to compute the Lyapunov table and (ii) Even if a model of the system were available, relatively small uncertainties and/or disturbances could provoke chaos in spite of the Lyapunov table indicates that chaos has been suppressed. The latter difficulty can be avoided by means of robust feedback control schemes. However, the designer of a feedback controller can stumble with the difficulty related with modeling errors [44]. For example, the problem of the synchronization of two strictly different chaotic oscillators requires control in spite of the differences between master and slave system [19], but such a difference are not always known.

Lemma 3.11. Let $x_i = x_{i,M} - x_{i,S}$ be the synchronization error for the master/slave interconnection in Eq. (3.8). Assume that the system $\dot{x} = f(x) + g(x)u$ (where $x \in \mathbb{R}^n$, $u \in \mathbb{R}$ while $f(x)$ and $g(x)$ are known vector fields) represents the chaotic dynamic of the synchronization error. In addition, suppose that the system output (measured state) is $y = Cx$, where C is a vector which can be chosen such that $y \in \mathbb{R}$. Then, the trajectories exponentially converges to zero if the feedback control force has perfect information of the dynamical system.

Proof. In seek of clarity and without loss of generality, consider the following feedback control $u = [-f(x) + V(x;k)]g^{-1}(x)$, where $V(x;k) = K^T x$ means the desired dynamics to be induced by the controller. Then, the closed-loop system (*i.e.*, the synchronization error system under control actions) becomes $\dot{x} = V(x;k)$. Since $V(x;k) = K^T x$ the control coefficients K are chosen such that the polynomial $s^\rho + k_1 s^{\rho-1} +\ldots+ k_{\rho-1}s + k_\rho = 0$ be Hurwitz. In this sense, such control constant represents the convergence rate. The above controller will exponentially steer the trajectories of the synchronization error system to zero at finite time t.

It is important to note that the perfect controller given by the feedback $u = [-f(x) + V(x;k)]g^{-1}(x)$ can be taken into the form (3.7). To this end, Taylor linearization can be applied to the feedback controller. After that, Laplace operator is used to get the resulting controller (3.7).

Figure 3.11 shows the dynamics of the synchronization error (all states of the synchronization error system are Exact Synchronized, ES). The master and slave oscillators are represented by the Lorenz equation. Note that the trajectories of the synchronization error system are leaded to zero. The synchronization was carried out by the ideal feedback: $u = [-f(x) + V(x;k)]g^{-1}(x)$ (perfect knowledge of $f(x)$). The controller was activated at $t = 50$ (*i.e.*, $u = 0$ for all $t < 50$). Note that such a feedback have prior knowledge about the nonlinear functions $f(x)$ and $g(x)$. In consequence, such controller is IFC [14]. The initial conditions of the slave system were chosen as follows: $x_S(0) = (-1.0,0.0,0.5)$. On the other hand, the initial condition for the Master system were chosen as: $x_M(0) = (10.0,-10.0,10.0)$. In this case the desired dynamics was chosen linear and is given by $V(x;k) = k_1x_1 + k_2x_2$, and the control constant values $k_1 = k_2 = 250.0$. Note that the trajectories of the synchronization error system converges exponentially to zero. Nevertheless, note that under the perfect controller (exact synchronization) the feedback requires large parameter values (high-gain feedback control).

Practical Synchronization

Definition 3.12. It is said that two chaotic systems are practically synchronized if the trajectories of the synchronization error $x_i = x_{i,M} - x_{i,S}$, for i = 1,2, ...,m, converge to a neighborhood around the origin. This implies that for all time $t \geq t^*$ the trajectories of the slave system are *close* to the master trajectories, *i.e.*, $x_{i,S} \approx x_{i,M}$.

Remark 3.13. Also Definition 3.12 involves open-loop and closed-loop control schemes. Several feedback control schemes have been proposed to lead chaotic trajectories around origin (see for instance [20], [34], [45]and [46]). Although for certain values of the control parameters the trajectories can be leaded almost to zero, the practical synchronization implies that $\| h(x_M, x_S) \| = \delta \cong 0$ holds along the trajectories of the synchronization error.

Lemma 3.14. Let $x_i = x_{i,M} - x_{i,S}$ be the synchronization error. Assume that the system $\dot{x} = f(x) + g(x)u$ (where $x \in \mathbb{R}^n$, $u \in \mathbb{R}$ while $f(x)$ and $g(x)$ are uncertain vector fields) represents the chaotic dynamic of the synchronization error. In addition, suppose that

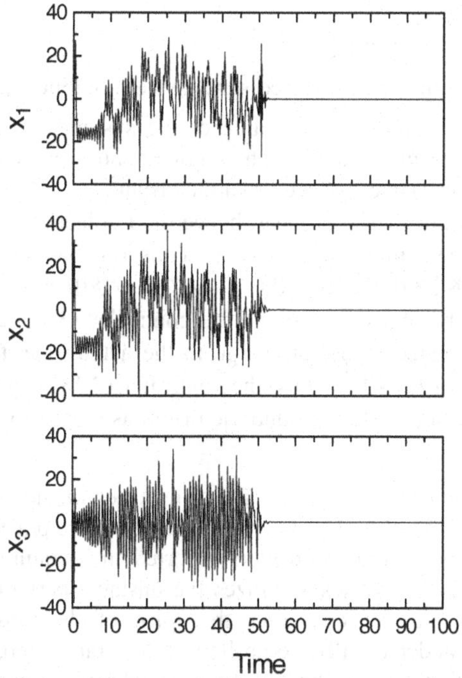

Fig. 3.11. Exact Synchronization. The dynamics of the synchronization error converges to zero. Note that exact synchronization corresponds to generalized synchronization for $x_S = Ix_M$ where I is the identity matrix.

the system output (measured state) is $y = Cx$, where C is a vector which can be chosen such that $y \in \mathbb{R}$. Then, the PII^2 controller yields the trajectories of the synchronization error system to an arbitrarily small neighborhood, which contains the origin.

Proof. Defining the coordinates exchange $z = \Phi(x)$, the system $x = f(x) + g(x)u$ can be rewritten as follows [25]

$$\begin{aligned} \dot{z}_i &= z_{i+1} \quad ; i = 1,2,\ldots,\rho - 1 \\ \dot{z}_\rho &= \alpha(z,v) + \gamma(z,v)u \\ \dot{v} &= \zeta(z,v) \end{aligned} \tag{3.19}$$

where $v \in \mathbb{R}^{n-\rho}$ represents the unobservable states, $\alpha(z,v)$ and $\gamma(z,v)$ are unknown nonlinear functions and ρ is the relative degree *i.e.*, the lowest order time-derivative such that the control command is directly related with the output.

Following the idea reported in [21] (also see [34] for high-order systems), one can construct a feedback which yields chaos control against the unknown functions $\alpha(z,v)$ and $\gamma(z,v)$. In particular, if $\rho = 1$ the feedback controller is given by the following equations

$$\begin{aligned} \dot{\hat{z}} &= \hat{\eta} + u + L\kappa_1(z - \hat{z}) \\ \dot{\hat{\eta}} &= L^2\kappa_2(z - \hat{z}) \\ u &= \left[\hat{\eta} + k(\hat{z} - z^*)\right] \end{aligned} \tag{3.20}$$

where $z = y \in \mathbb{R}$ represents the measured state (system output) of the synchronization error system, $\hat{\mathbf{z}} \in \mathbb{R}$ is an estimated value of the system output, $\hat{\eta} \in \mathbb{R}$ provides an estimated value of the uncertainties. The unique control parameter is denoted by L (which is positive defined) and k, κ_1 and κ_2 are constant.

Note that the controller (3.18) is linear, hence the Laplace operator can be used for its transformation to the equivalent form (3.7). Thus, one has the following equations $s\hat{z}(s) = \hat{\eta}(s) + u(s) + L\kappa_1[z(s) - \hat{z}(s)]$; $s\hat{\eta}(s) = L^2\kappa_2[z(s) - \hat{z}(s)]$; $u(s) = \hat{\eta}(s) + k[z(s) - \hat{z}(s)]$. Then, by combining the above equations, the controller (3.7) can be obtained. The idea is that if the estimated values converge to the actual one, the controller (3.18) recovers the structure of the PII^2. Thus, the controller (3.18) can lead the trajectories of a given synchronization system around the origin as closed loop stability is asured with controller (3.7) [21]. □

Remark 3.15. The controller (3.7) leads the trajectories of the synchronization error system around zero, this implies that the master and slave are practically synchronized, *i.e.*, $x_{i,S} \approx x_{i,M}$. Perhaps several controllers reported in literature can yield practical synchronization. The PII^2 feedback stabilizes the unstable periodic orbits (UPO's) in this sense, the PII^2 controller conserves the spirit of the most proposed control schemes. In addition, under the PII^2 controller, an arbitrary reference can be tracked. Notice that the Definition 3.12 implies the phase synchronization is locking, *i.e.*, if the trajectories of the slave system are close of the master, then their phases are similar (see Figure 3.12).

Definition 3.16. It is said that two chaotic systems are completely synchronized if and only if all states of both master and slave systems are practically or exactly synchronized.

Figure 3.12 shows the dynamics of the synchronization error for the case of the Complete Practical Synchronization (CPS) of the Lorenz equation. The parameters values of the master system were chosen as follows: $\sigma_M = 11.0$, $r_M = 27$ and $b_M = 2.57$. The slave parameters values were chosen as follows: $\sigma_S = 10.0$, $r_S = 28.0$ and $b_S = 8/3$. The initial conditions were chosen as follows: $x_M(0) = (1.0,1.0,10.0)$ and $x_S(0) = (-1.0,10.0\text{-}1.0)$. The controller was activated at t = 50 (this is, for $t \leq 50$, $u = 0$). The control parameter values were chosen as $k_C = k_e = 8.0$ and the high gain parameter L = 10. The controller (3.7) steers the trajectories of the synchronization error around zero. This means that the synchronization is not exact. The phase of the synchronization error is locking. However, the amplitude of the slave system is not the same than the master. In fact, the difference between master and slave does not display a regular behavior.

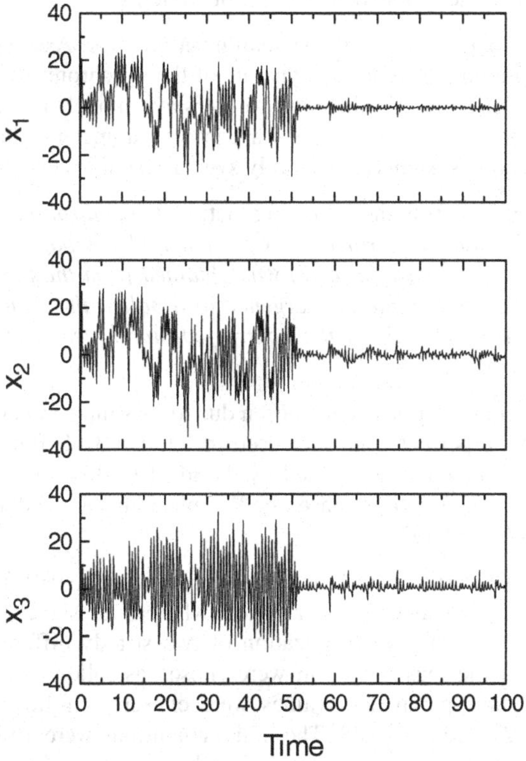

Fig. 3.12. Complete Practical Synchronization. The dynamics of the synchronization error converges around the origin.

In this sense, the Definition 3.12 agrees with the previous results reported in [29]. Moreover, such definition provides a geometrical notion of this kind of synchronization (see below).

Partial-state Synchronization

Definition 3.17. It is said that two chaotic systems are partially synchronized in states if, at least, one of the states of the slave system is synchronized with the corresponding in the master system or only if, at least, one of the states does not synchronize.

Remark 3.18. Partial-state synchronization has been found in several systems. In particular, partial-state synchronization can be found in networks of identical oscillators (even in the absence of noise). Indeed, some results has been published where the open-loop partial state synchronized has been studied [37], which contains the first evidence that the partial state synchronization can be related with the bifurcation parameter of the synchronization error system. Nevertheless, the definition of the partial state synchronization is assumed. Definition 3.17 is general and includes open-loop as well as closed-loop synchronization systems.

Lemma 3.19. Let $x_i = x_{i,M} - x_{i,S}$ be the synchronization error. Assume that the system $x = f(x) + g(x)u$ represents the chaotic dynamic of the synchronization error. Suppose that the master model is strictly different to the slave model. Besides, consider the feedback control (3.7). Then, there exists a control parameter value such that both master and slave chaotic systems are partially synchronized in states.

Proof. The proof follows from the following fact: *As the control parameters increase, the controller (3.7) leads the trajectories to a neighborhood, which contains the origin. This is, since the controller (3.7) was obtained from the feedback (3.18) and such controller yields asymptotic convergence to a ball with radius* $r = O(L^{-1})$ *(for more details, see* [26]*). Hence, as the control constants decrease as the complete synchronization is lost.* □

Remark 3.20. PII^2 controller has been chosen due to its simple structure. In fact it is a robust approach to the perfect controller used in Lemma 3.11. Finally, note that the partial synchronization cannot be yielded by the ideal feedback due to it cancels the nonlinearities $f(x)$ and $g(x)$. In this sense, the PsS phenomena is induced by the robust features of the feedback (3.18).

Figure 3.13 shows the Partial state Practical Synchronization (PsPS). In this case, the Lorenz system was chosen as the master system whereas the slave system is given by the Chua oscillator (*i.e.*, the synchronization of two strictly different systems). The parameters values of the master system were chosen as follows: $\sigma = 10.0$, $r = 28.0$ and $b = 8/3$. The slave parameters values were chosen as follows: $g_1 = 10.0$ and $g_2 = -14.87$, $\alpha = -1.27$ and $\beta = -0.68$. The initial conditions were arbitrarily chosen as follows: $x_{1,M} = x_{1,S} = 0.074$, $x_{2,M} = x_{2S} = -0.023$ and $x_{1,M} = x_{1,S} = -0.063$.

A schematic representation of the geometrical interpretation of the Exact Complete Synchronization is illustrated in the Figure 3.14a. In such case the trajectories of the synchronization error converges to origin. Figure 3.14b shows the geometrical

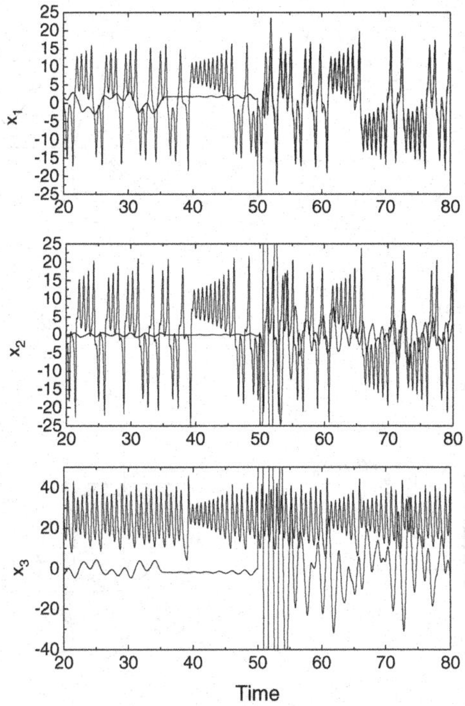

Fig. 3.13. Partial-state Practical Synchronization. The dynamics of, at least, one state of the synchronization error system is synchronization while, at least, another one is not synchronous. The thin-line is the dynamics of the slave system.

interpretation of the Practical Complete synchronization, Here, the synchronization error dynamics converges to a ball (represented by a *sphere*), which contains the origin. On the other hand, Figure 3.14c shows the schematic interpretation of the Partial state Practical Synchronization. In this case, the trajectories are leaded to a ball. The ball has been represented by a *cylinder* (if only one state of the error system is not synchronized) or by a *sheet* (if two states of the error system are not synchronized, see Figure 3.14d). The idea behind the interpretation is as follows: *If the complete synchronization is lost, then the ball is deformed in any direction.*

The controller was chosen as the feedback (3.7) and the control parameters values were chosen as follows: $\tau_I = \tau_{II} = 1.0$, $k_c = k_e = 15.0$. The controller was activated at $t = 50$ (*i.e.*, $u = 0$ for $t \leq 50$). In such case, the first state, $x_1 = x_{1,M} - x_{1,S}$, has been synchronized. Note that the dynamics of the second state ($x_2 = x_{2,M} - x_{2,S}$) and the third one ($x_3 = x_{3,M} - x_{3,S}$) are not synchronous.

Almost Synchronization

The controller (3.7) is able to suppress the chaotic behavior of the synchronization error system. In this way the synchronization of both master and slave system can be achieved. However, the control constants, k_c, k_e, τ_I and τ_{II}, can be interpreted as

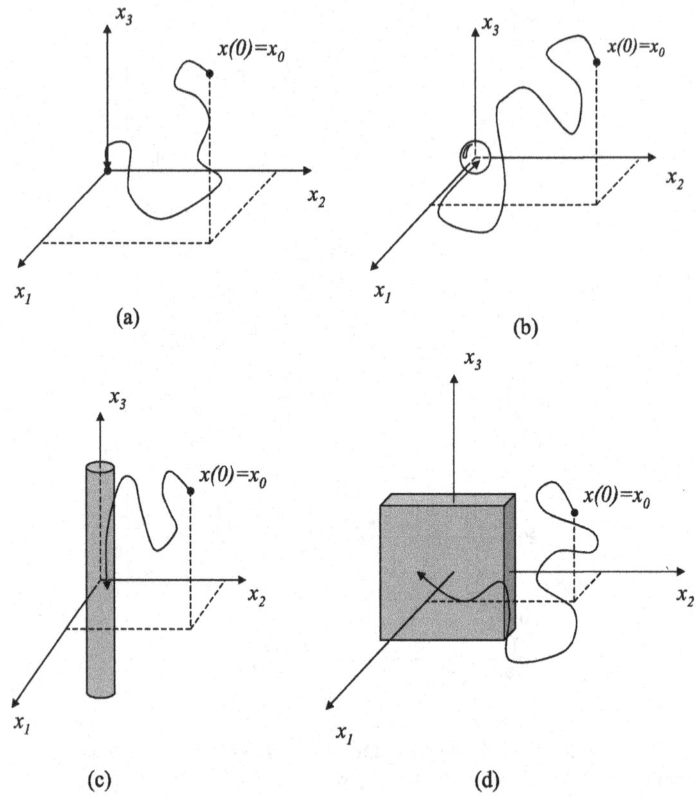

Fig. 3.14. Geometrical interpretation of the synchronization phenomena. (a) Exact Synchronization, (b) Practical Synchronization, (c) and (d) Partial-state synchronization.

bifurcation parameters (due to for different values of such parameters one can found different dynamical behavior of the synchronization error system). This is, if the value of the control parameters change the synchronization can be lost into some states of the slave system. Even if the master and slave system are strictly different, the controller (3.7) is able to carry out the chaos synchronization. This is due to the controller was designed in such way that, in spite of the model errors, the controller can stabilize the synchronization error trajectories around the origin.

Synchronization of nonlinear systems is a very important and interesting problem, which is being widely studied. On the contrary, asynchronous states have received less attention (a few papers can be found in the literature, for instance see [47] and references therein). Nevertheless, such phenomenon is very important and interesting. For example, in a healthy piece of brain tissue, asynchronous states of the neurons (which can be modeled as coupled nonlinear oscillators) can be characteristic of epileptic activity [48]. Here, it is shown, via numerical simulations, that nonlinear systems can display synchronous as well as asynchronous states.

Definition 3.21. It is said that two chaotic systems are almost synchronized if at least one state in both master and slave systems displays oscillations with the same phase and different amplitude and another one is not synchronized.

Remark 3.22. Notice that the above definition does not exclude partial state synchronization. This is, in principle, it is possible to find the Almost Partial-state Synchronization (APS). Indeed, the APS can be induced by the controller (3.7). Figure 3.15 shows the synchronization of the Lorenz system and the Chua oscillator. The Lorenz system was chosen as master whereas the Chua oscillator becomes the slave system. The parameter values of the Lorenz system were chosen as follows: $\sigma = 10.0$, $r = 28.0$ and $b = 8/3$. The parameter values of the Chua oscillator were selected as: $g_1 = 10.0$, $g_2 = -14.87$ $\alpha = -1.27$ and β 0 -0.68. The initial conditions were arbitrarily chosen as follows: $x_{1,M} = x_{1,S} = 0.074$, $x_{2,M} = x_{2,S} = -0.023$ and $x_{3,M} = x_{3.S} = -0.063$. Note that the state $x_1 = x_{1,M} - x_{1,S}$ is practically synchronized while the state $x_2 = x_{2,M} - x_{2,S}$ is almost synchronized and the state $x_3 = x_{3,M} - x_{3,S}$ is not in synchrony.

Here, we have briefly analysed the synchronization phenomena. The synchronization phenomena were characterized in five types: (i) Exact, (ii) Practical, (iii) Complete, (iv) Partial-state and (v) Almost synchronization. To this end a feedback control structure was used. The feedback structure holds the main features of several chaos control schemes. In addition, in order to illustrate each synchronization type numerical simulations were performed. Each kind of synchronization has been previously reported. For example, the practical synchronization has been reported in [29] and [37].

Thus in this section the evidences for some synchronization phenomena were showed. Moreover definitions for the distinct phenomena were stated. Such definition are based on the geometrical and structural features where the synchrony is fundamental. Moreover, the characteristics of the synchronization phenomena were established. In addition, the main differences between the synchronization types were studied. Exact, Practical, Partial state and Complete synchronization can be found if the master and slave model have the same structure. Finally, this section shows that it is possible to find that the combination of the synchronization phenomena can be displayed by the same system. Thus, the main factors are the control parameters.

Conjecture 3.23. The almost synchronization can be found if (i) Both master and slave systems are strictly different, and (ii) The control structure is based on feedback with lack knowledge about the master and slave.

Exercise 3.24. *Application of ES on communication.* Consider two identical chaotic systems given by eq. (2.3). Then consider an input signal $u_M = \alpha \sin(\omega t)$ injected by the vector $g_M = [1\ 0\ 0]^T$ to the master, with $\alpha = 5$ and $\omega = 5$; then, find the control command u_S injected by the vector $g_S = [1\ 0\ 0]^T$ to the slave system for $t = t_0 \neq 0$ such that Exact Synchronization between them appears. Finally, shows that the control signal u_S is equal to the transmitted signal once both chaotic systems have been synchronized. Hint: Use the result given in Lemma 3.8.

Exercise 3.25. *Application of CPS on communication.* Repeat Exercise 3.24 but with two chaotic systems (eq. 2.3) with same model but with a small difference parameter (for instance $\sigma = 10.5$ in the slave system). Then shows that the control signal u_S is close to the transmitted signal u_M.

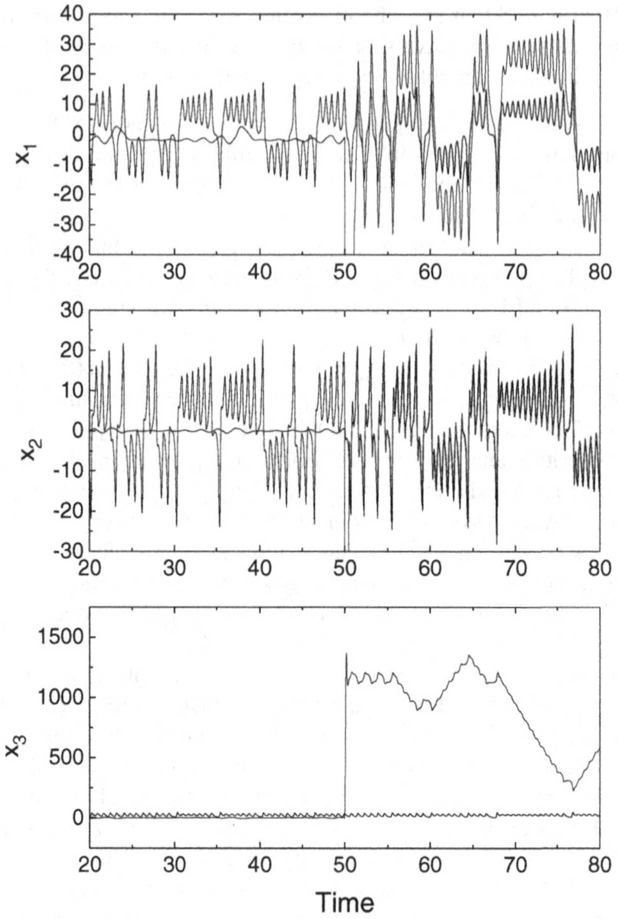

Fig. 3.15. Almost Practical Synchronization. One state is practically synchronized while the second state has the same phase than the corresponding master state. Finally, the last one is not synchronous.

3.3 Chaos-Based Signal Transmission Despite Master/Slave Mismatches: An Experimental Test

A standard approach of the spread-spectrum communications involves the addition of a message signal to a carrier at the transmitter in such way that, at the receiver system, the carrier signal is removed to reveal the message. As a consequence, the message signal can be unveiled. The use of chaotic carriers instead of random carrier produces similar spreading of the spectrum due to broad-band nature of the chaotic carrier. Consequently, the use of spectral analysis techniques to reveal the message will generally fail. In order to remove the carrier signal, the synchronization of the carrier with the receiver is desired. However, synchronization is not an easy task if we think that: (a) due to sensitive dependence of chaos on initial conditions. If initial conditions of the carrier are not known then the trajectories of the transmitter cannot be recovered (b) in matching exactly the drive and response systems, even infinitesimal parametric difference of any model will eventually result in divergence of orbits starting nearby each other. That is, the goal is to synchronize the receiver with transmitter by means of a designed coupling force u in such way that the trajectories of the receiver (response system) tracks the trajectories of the carrier (drive system) in spite of the parameter mismatches and uncertainties.

In spite of above difficulties, application of the chaos synchronization in secure communication has been intensively studied in last decade. The underlying idea behind the chaos-based communication is to use a carrier (which is given by a chaotic system) as a type of spread-spectrum communication system. To separate the hidden message from a chaotic signal is not an easy task [49]; hence chaos-based communication has potential application in secure communication. As a matter of fact, several chaos-based communication schemes have been published (see, for instance, [1], [50] [51] and [52]). Earlier 90's two basic configurations were identified: (i) An approach consists of the addition of the message signal to the chaotic carrier. Then, the transmitted signal is sent to the receiver, which is a reduced model of the carrier. That is, transmitter comprises the full-state model whereas receiver is composed by a reduced model [1], [50] and [51]. (ii) Another transmitter/receiver design is based on the full-state model of the driving and response systems. That is, both drive and response systems are represented by dynamical systems of the same order [1],[11]. The above configurations have been addressed via several synchronization schemes. For example: (A) the homogeneous synchronization configuration has been recently addressed via parameter modulation. Kocarev and Parlitz [18] have proposed a generalization of these approaches which extends the capabilities for constructing synchronized systems. Their approach enables the message signal to be integrated as a driving signal. However, the message signal can be recovered only under ideal conditions; (B) the monotone synchronization is a feedback scheme based on high-gain feedback [14]. The monotone synchronization has application to secure communication; however, it is based on high-gain feedback which is very sensitive to noise; (C) adaptive synchronization was initially reported in [15] and, potential application of adaptive synchronization have been recently reported [52]; and finally, (D) a theoretical framework was reported in [15] with promissory results and potential application to secure communication. Such results are related with adaptive synchronization, however the theoretical framework is based on geometrical control theory [25]. The goal is to stabilize, at the origin of the synchronization error system, i.e., the discrepancy between the driving and response system. The discrepancy is defined as the dynamical differences between driving and response system and includes: (i) model mismatches, which means that the model of the drive system could not be the same than the response system. (ii) Unknown initial conditions, which implies that the time series of the transmitter cannot be equal than those of the receiver. (iii) Parametric uncertainty, which means that the receiver devices could be constructed with inaccuracies. By using such a theoretical framework, an experimental application is presented. The main idea is, departing from the system that represents the discrepancies, construct an extended nonlinear system which should be dynamically equivalent to the canonical representation. In this way, the differing terms are lumped in a nonlinear function, which is rewritten into the extended nonlinear system as a state variable. Thus an observer can be constructed to get an estimated value of the lumping nonlinear function via the augmented state variable.

In order to clarify the previous discussion an experiment was carried out communicating two computers. The Chua circuit was simulated numerically in both transmitter and receiver computers. Parameters and initial conditions were assumed different. Delay in transmission was neglected. Although transmission was performed in noisy environment, the message signal was acceptably recovered. In fact, a performance of the synchronization scheme was measured by an index. The experiments were performed by means of MatLab/Simulink software.

3.3.1 The Chaos-Based Scheme

Let us consider two third-order chaotic oscillators, which will be used as drive and response system. In particular, let us assume that both drive and response systems are given by Chua circuit, which yields a double-scroll attractor. Suppose that modulated and demodulated signals are carried out by the same channel, i.e., $g_S(x_S) = g_M(x_M) = (1,0,0)^T$. Thus, the drive system can be written in dimensionless form as follows

$$\begin{aligned} \dot{x}_{1,M} &= \pi_{1,M}[x_{2,M} - x_{1,M} - f(x_{1,M})] + \lambda_m \\ \dot{x}_{2,M} &= x_{1,M} - x_{2,M} + x_{3,M} \\ \dot{x}_{3,M} &= -\pi_{2,M}x_{2,M} \end{aligned} \quad (3.21)$$

$f(x_{1,M}) = \pi_{3,M}x_{1,M} + ½(\pi_{4,M} - \pi_{3,M})(|x_{1,M} + 1| - |x_{1,M} - 1|)$, λ_m is the transmitted signal. Suppose that the same configuration is used as a response system. However, assume that there are differences between resistance). That is, the parameter values of the receiver are different than the transmitter. In this way the response system becomes

$$\begin{aligned} \dot{x}_{1,S} &= \pi_{1,S}[x_{2,S} - x_{1,S} - f(x_{1,S})] + u \\ \dot{x}_{2,S} &= x_{1,S} - x_{2,S} + x_{3,S} \\ \dot{x}_{3,S} &= -\pi_{2,S}x_{2,S} \end{aligned} \quad (3.22)$$

$f(x_{1,S}) = \pi_{3,S}x_{1,S} + ½(\pi_{4,S} - \pi_{3,S})(|x_{1,S} + 1| - |x_{1,S} - 1|)$, with $\pi_S \neq \pi_M$. From the differences $x_i = x_{i,M} - x_{i,S}$, one has that the system of the following synchronization discrepancy can be obtained: $\dot{x}_1 = \Delta f_1(x,x_M) + \lambda_m - u$, $\dot{x}_2 = \Delta f_2(x,x_M)$, $\dot{x}_3 = \Delta f_3(x,x_M)$, with $\Delta f_1(x,x_M) = \pi_{1,M}[x_{2,M} - x_{1,M} - f(x_{1,M})] - \pi_{1,S}[x_{2,S} - x_{1,S} - f(x_{1,S})]$, $\Delta f_2(x,x_M) = (x_{1,M} - x_{2,M} + x_{3,M}) - (x_{1,S} - x_{2,S} + x_{3,S})$, $\Delta f_3(x,x_M) = -\pi_{2,M}x_{2,M} + \pi_{2,S}x_{2,S}$ and $x_{i,S} = x_i - x_{i,M}$. Or equivalently,

$$\dot{x} = \Delta f(x,x_M;\Delta\pi) + g(x)(\lambda_m - u) \quad (3.23)$$

where Δf: $\mathbb{R}^3 \rightarrow \mathbb{R}^3$ is a smooth map, $\Delta\pi$ denotes the differences between parameters π_M and π_S, $g(x) = [1,0,0]^T$ is the modulation/demodulation channel. System (3.21) is obviously uncertain. That is, since difference between systems (3.19) and (3.20) is unknown, the map Δf: $\mathbb{R}^3 \rightarrow \mathbb{R}^3$ is unknown. However, note that, if system (3.21) can be stabilized at origin for any time $t \geq 0$, then differences between systems (3.19) and (3.20) tends to zero. That is, $\Delta f(x,x_M;\Delta\pi)$ holds close to zero and, consequently, the coupling function u is close to the message signal λ_m. Note that λ_m is also unknown.

Thus, the aim is to design a control u such that the trajectories $x(t) = x_M(t) - x_S(t) \rightarrow 0$ for all time $t \geq t_0 \geq 0$ (where t_0 is the time of control activation) and any initial conditions $x(0) = x_M(0) - x_S(0) \in \mathbb{R}^3$ in spite of parametric uncertainties $\Delta\pi$. In order to address the stabilization of the uncertain system (3.21), let us assume that:

A.1) The vector field $\Delta f(x, x_M; \Delta\pi)$ is smooth but uncertain
A.2) The vector field $g(x)$ is known and bounded away from zero
A.3) The scalar function $y = h(x) = x_{1,M} - x_{1,S}$ is the system output, where $x_{1,M}$ is the transmitted signal and $x_{1,S}$ is the corresponding signal into the receiver.

By taking $y_M = x_{1,M}$ and the receiver output by $y_S = x_{1,S}$, one has that the output of discrepancy system becomes $h(x) = x_1$. This implies that the smallest integer such that: (i) $L_g L_f^i \boldsymbol{h}(\boldsymbol{x}) = 0$, $i = 1,2,\ldots, \rho-2$ and (ii) $L_g L_f^\rho \boldsymbol{h}(\boldsymbol{x}) \neq 0$ is $\rho = 1$ [53]. In this way there is a coordinate transformation which is globally defined and becomes [53]: $z_1 = x_1$, $v_1 = x_2$ and $v_2 = x_3$. Then the discrepancy system can be rewritten as

$$\begin{aligned} \dot{z}_1 &= \Delta f_1(z,v) + \lambda_m - u \\ \dot{v}_1 &= \Delta f_2(z,v) \\ \dot{v}_2 &= \Delta f_3(z,v) \\ y &= z_1 = x_1 \end{aligned} \tag{3.22}$$

where $\Delta f_i(z,v) = f_1(x_M) - f_1(x_S)$ is an unknown function and y denotes the output of the discrepancy system. According to results in [53], system (3.22) is minimum phase (i.e., the subsystem $= \zeta(z,v)$ is stable, where $\zeta(z,v) = (\Delta f_2(z,v), \Delta f_3(z,v))^T$). This implies that there is a feedback control such that synchronization is achieved and message signal can be recovered [53]. It should be noted that states $v \in \mathbb{R}^{n-\rho}$ of the subsystem $= \zeta(z,v)$ are obtained following the procedure in Chapter 4 of [25].

Now, since $\Delta f_1(z,v)$ and λ_m are unknown, one can define the following $\eta = \Delta f_1(z,v) + \lambda_m$. Then, system (3.21) becomes

$$\begin{aligned} \dot{z}_1 &= \eta + u \\ \dot{\eta} &= \Gamma(z, v, u, \lambda_m) \\ \dot{v}_1 &= \Delta f_2(z,v) \\ \dot{v}_2 &= \Delta f_3(z,v) \\ y &= z_1 \end{aligned} \tag{3.25}$$

where, according to Assumption A.1, $\Gamma(z,v,u,\lambda_m)$ is an unknown function. Moreover, by definition, $\eta = \Delta f_1(z,v) + \lambda_m$ is not available for feedback. Therefore, only $y = x_{1,M} - x_{1,S}$ is available for feedback.

Following ideas in [25] and [54], one feedback controller can be designed to stabilize system (3.23) at origin, and as a consequence system (3.22). Thus, the proposed feedback is given by

$$\begin{aligned} \dot{\hat{z}}_1 &= \hat{\eta} - u + L\kappa_1(z_1 - \hat{z}_1) \\ \dot{\hat{\eta}} &= L^2\kappa_2(z_1 - \hat{z}_1) \\ u &= \hat{\eta} + k_1\hat{z}_1 \end{aligned} \tag{3.26}$$

where $(\hat{z},\hat{\eta})$ denotes the estimated values of the states (z,η), (compare eq. (2.12) in Theorem 2.8 and remark 2.9 with eq. (3.24)). Note that controller (3.24) does not require a priori information about the carrier model. In fact, the idea behind the proposed controller is to obtain estimates values $(\hat{z}(t),\hat{\eta}(t))$ very close to the actual values $(z(t),\eta(t))$ for all time $t \geq t_0 \geq 0$, where t_0 is the time of the control activation. That is, if $(\hat{z}(t), \hat{\eta}(t)) \approx (z(t),\eta(t)) = ((x_{1,M}(t) - x_{1,S}(t)),(\Delta f_1(z,v) + \lambda_m))$ for all $t \geq t_0 \geq 0$, the control command becomes $u \approx \Delta f_1(z,v) + \lambda_m + k_1 z_1$. Hence, if $(\hat{z}(t), \hat{\eta}(t)) = (z(t),\eta(t))$, system (3.22) under the proposed feedback becomes: $\dot{z}_1 = k_1 z_1$, $\dot{v}_1 = \Delta f_2(z,v)$ and $\dot{v}_2 = \Delta f_3(z,v)$. Since, subsystem $\dot{v} = \zeta(z,v)$ is stable, the trajectories of the closed-loop system converge to origin. Note that $(z_1,v_1,v_2) = (0,0,0)$ for al $t \geq t_0 \geq 0$ implies that $x_S(t) = x_M(t)$ for all $t \geq t_0 \geq 0$, from where $\Delta f_i(x,x_M) = f_i(x_S) - f_i(x_M) = 0$ and $g(x)(\lambda_m - u) = 0$. Hence, since $|\, g(x) \,| > 0$, as (z_1,v_1,v_2) tends to zero as the difference $\lambda_m - u$ tends to zero. That is, the coupling between drive system (3.19) and response system (3.20) recovers the message signal.

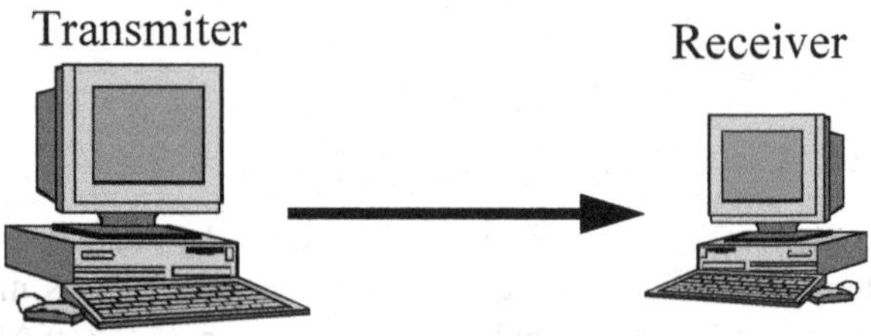

Fig. 3.16. The experiments were carried out in laboratory from the communication of two computers. The transmitter and receiver were numerically simulated in both computers. In transmitter the message signal was entered to the system (3.19) whereas receiver includes system (3.20) and controller (3.24).

3.3.2 Experimental Setup

Experimental signal transmission from synchronization of the Chua circuit has several alternative procedures. For instance: (i) a transmission scheme was reported in [55], which is based on neural networks. Results in [55] show that neural-networks-based circuits, generating the dynamics of the Chua oscillator, can be used for signal transmission with potential application and (ii) Torres and Aguirre [56] have reported an architecture for Chua circuit with operational amplifier realization of the inductor. Such a circuit has broadband spectrum, which can be concentrated at low-frequencies. Thus, the "inductorless " circuit can be used for chaos-based transmission proposals.

In this example two computers were communicated for testing the synchronization scheme, see Figure 3.16. The transmitter is given by system (3.19) and is numerically simulated in first computer while message signal was entered in the oscillator. Then $y = x_1$ is sent via a commercial communication system, which is connected from parallel port. The receiver is given by system (3.20) and is numerically simulated in second computer while controller (3.24) is included. The following control parameter values were chosen: (control gain) $k_1 = 7.0$, (estimation constants) κ_1 and κ_2, were chosen in such way that the polynomial $s^2 + \kappa_1 s + \kappa_2 = 0$ has its roots at -1.0 and (high- gain estimation parameter) L = 100. The initial conditions were chosen as follows: $x_M(0) = (0.01,-0.03,0.02)$, $x_S(0) = (0.02,-0.04,-0.01)$ and $(\hat{z}_1(0),\hat{\eta}(0)) = (-0.02,-0.01)$. To carry out the experiments, nominal values of the transmitter parameter and receiver were chosen equal, i.e., $\pi^*S = \pi^*M$, where star denotes the nominal value. Parametric variations were induced into the receiver, i.e., $\pi_S = \pi^*M(1 + \%\Delta)$, where $\%\Delta$ is the percentage of parametric variation. The message signal was arbitrarily chosen as $\lambda = 0.2\sin(0.63t)$. The unique criteria was that amplitude and frequency were such that transmitter remains in chaotic behavior. Similar results were obtained for digital message.

Figure 3.17 shows the power spectrum of the transmitted signal y and recovered signal u for several parametric differences $\%\Delta$ at the receiver transmitter $\pi_{1,S}$. Note that frequency of the message signal is $\phi \approx 0.11$ and compare with power spectrum of the recovered signal (Figs. 3.17b - 17d), That is, if the parameter values in transmitter and receiver are identical, the message signal is recovered exactly, i.e., without noise. On the contrary, if parametric differences are not zero, the message signal is recovered and a filtering step is required. In addition, note that frequency of message signal is *masked* into the transmitted signal (see Figure 3.17a).Note that parameter $\pi_{1,S}$ is, in some sense, related with observable state z. However, parameter $\pi_{2,S}$ is related

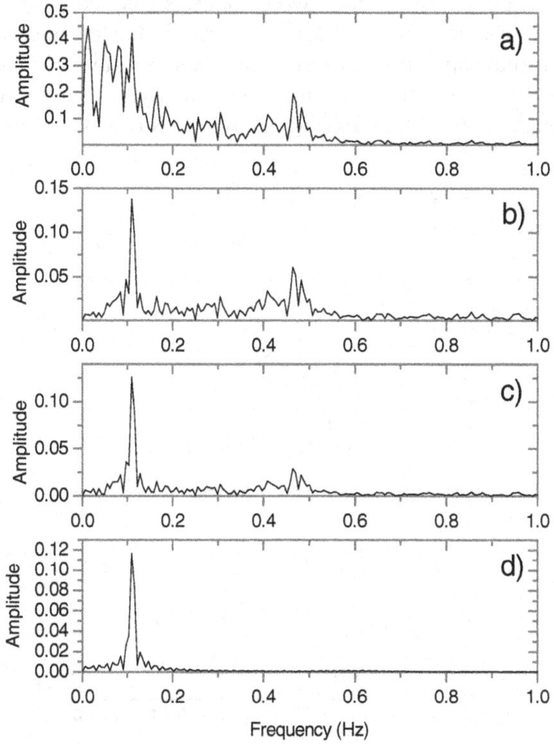

Fig. 3.17. Power spectrum of: (a) transmitted signal $y = x_{1,M}$ and (b), (c) and (d) the recovered signal u under the following parametric differences $\pi_{1,S}$-10%,-5% and 0%, respectively. Frequency of message signal λ_m is ϕ =0.11 Hz.

with unobservable states (v_1,v_2). Figure 3.17 shows the power spectrum of signals for several parametric differences %Δ at $\pi_{2,S}$. Same results were obtained for these experiments; i.e., system (3.24) provides stability margin in recovering the message signal against parametric variations. It should be pointed out that similar results were obtained for variations in parameter $\pi_{2,M}$. Figures 3.17 provides information about the frequencies into the transmitted and recovered signals. However, it is desirable to get a quantitative measure of recovering.

The following performance index was defined to get a quantitative measure of the recovering accuracy: $I_P = \int^t t \mid \lambda(\sigma) - u(\sigma) \mid d\sigma$, where t denotes the experiment time. Such index, I_P, allows to measure the performance for long time, (t >> 1), diminishing the effect of transients. In this sense, I_P is a measure for long time transmissions. Two set of experiments were carried out to test the controller (3.24).

Figure 3.18 shows the performance index *versus* percentage of parametric variation. As the parametric differences decreases as recovering accuracy increases. In this sense, the message signal can be recovered in spite of transmitter/receiver mismatches.

The proposed scheme departs from geometrical control theory and has adaptive structure which only comprises two tuning parameters: The high-gain estimation L and the control gain k. A performance index is used to indicate that promissory results are obtained in laboratory. To this end, two computers were communicated. Since the drive and response systems are nonlinear, the transmitter/receiver mismatches can yield additive (or multiplicative) uncertainties. Such results is a consequence of synchronization of the nonlinear oscillators. Actually, a central question is the following: How stable is the synchronized behavior?. Since synchronization has been

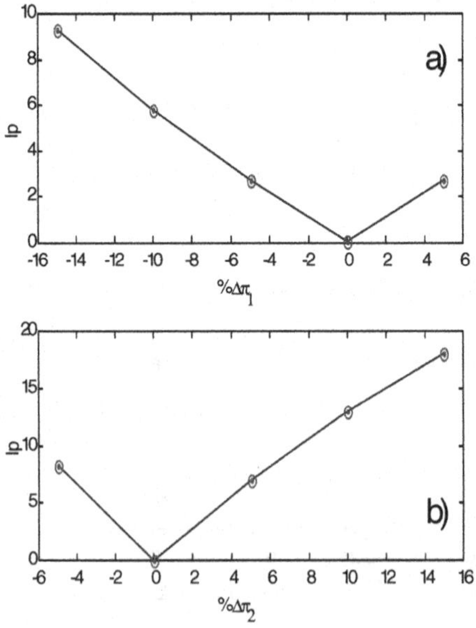

Fig. 3.18. Performance index I_p for several percentage variation. (a) Variations in $\pi_{1,S}$, (b) variation in $\pi_{2,S}$.

addressed as a stabilization problem, robust stability of parametric uncertainties can be dealt via H_∞ control theory [7], [17]. In next section a local robust analysis of chaos synchronization is performed.

3.4 Robust Stability Analysis of Chaos Synchronization

Essential elements on chaos synchronization problem are: (i) Slave system trajectories should track the master system trajectories and (ii) synchronization remains against uncertainties due to differences between oscillators. Chaotic synchronization is the regime in which two chaotic systems exhibit identical oscillations, i.e., synchronization error tends to zero for all $t \geq 0$. Chaos synchronization seems difficult task; over all if we think that: (a) due to sensitive dependence of chaotic flows on initial conditions, (b) in matching exactly the master and slave systems, (c) parametric differences between chaotic systems (for instance, due to inaccuracy design or time-variations) yield different attractors. Synchronization is not an easy task, since 90's several synchronization schemes have been proposed [14], [15], [19], [35], [57]. In fact, the synchronization of dynamical systems against unknown initial conditions is solved [14],[35]. Regarding parameter uncertainties, several alternative synchronization procedures can be found in literature. For example, some synchronization results are the following: Adaptive feedback schemes have been reported (see [34], [57] and references therein) and high-gain synchronization was discussed in [1] and [22]. Also, synchronization has been interpreted as an observer problem [58]. On the other hand, parameter modulation and noise effect have been also studied [18], [59]. Moreover, some feedback schemes have been reported under which synchronization is achieved against strictly different model [19], [33]. In addition, several kinds of synchronization can be displayed for chaotic systems under feedback control [5], [29].

A central question in theory on synchronized chaotic systems is the following: *When is the origin of synchronization error system stable?. That is, how stable is the synchronized behavior?.* The notion of robust synchronization has been recently discussed in [46], [60] and [61]. Robustness of the chaos synchronization is related with transversal stability theory, which is based on Lyapunov's stability theory. Robust synchronization has been studied from two perspectives. On the one hand, authors in [46] reported a procedure to measure the divergence between orbits of the synchronization system. That is two synchronized systems behaves in similar manner, hence orbits of both response and drive systems will be close for all time $t \geq t_0 \geq 0$. Since Lyapunov exponents provide a divergence measure of each orbit in a chaotic attractor, authors in [46] used them as a measure of the divergence size of the synchronized systems. However, although Lyapunov exponents provides a global measure, some arbitrary factors are required to compute the largest Lyapunov exponent. For example, length of trajectory segment along which the Lyapunov exponents are computed. On the other hand, authors in [60] and [61] studied robust synchronization by minimizing the L_2-gain from the exogenous input to the system of the drive/response discrepancy (see [62] for an introduction). L_2-gain techniques provides and algorithm for designing a synchronization scheme. The L_2-gain problem is *to design a control input in order to minimized the difference $y - r$, which tends to zero for all $t \geq t_0 \geq 0$ (where y denotes the output of the controlled system whereas r*

denotes the reference). That is, the L_2-gain problem is an optimal control problem. However, by analogy with the problem of the stabilization of the linear-time-invariant systems (which is so-called H_∞-control, see [17] or [44]), such a problem is so-called nonlinear H_∞control problem. The nonlinear H_∞ synchronization is a nonlinear H_∞ control problem and has direct application in secure communication.

We are interested in to study robust stability of the chaotic synchronization. That is, we wish to derive a measure of the stability of the synchronized systems. However, the procedure does not require the computation of the Lyapunov exponents. Robust stability notions have been developed from H_∞control theory [17],[44]. Such engineering methods are also related to transversal stability. However, H_∞control theory deals with robustness margins around a point of state space whereas Lyapunov exponents deals with divergence of the synchronization orbits. For synchronization case, the interesting point is, obviously, zero synchronization error, (i.e., $x(t) := x_M(t) - x_S(t) = 0$ for all $t \geq t_0 \geq 0$, where t_0 is the time of the control activation). This means that, from the control theory point of view, robust synchronization margin should be analysed into state space synchronization error. As we will see below, the physical interpretation of robust synchronization margin is related with the size of a neighborhood containing the origin, i.e., size of subset containing origin where master/slave discrepancy is evolving for all $t \geq t_0 \geq 0$. To this end, synchronization problem is solved via stabilization of synchronization error, which means that controller asymptotically steers the synchronization error trajectories $x(t) = x_M(t) - x_S(t)$ around the origin for all $t \geq t_0 \geq 0$. A robust stability measure is obtained from synchronized system using the H_∞-norm. According to robust control theory, the results imply that there is a control parameter value such that a large robustness margin can be achieved in complete practical synchronization. The robustness margin is related with singular values. Since synchronization is seem as the stabilization of the error $x(t) = x_M(t) - x_S(t)$ around the point $x^* = 0$, hence singular values provide a comparable robustness measure than Lyapunov exponents in sense of transversal stability of the synchronization system.

3.4.1 Chaos Synchronization Via Linear Feedback

Let us consider chaotic systems with the following structure

$$\dot{x}_M = f_M(x_M; p_1) + \tau_M(t) \tag{3.27}$$

$$\dot{x}_S = f_S(x_S; p_2) + \tau_S(t) + Bu \tag{3.28}$$

where subscripts M and S denotes master and slave system, respectively, $x_k \in \mathbb{R}^2$ is a state vector, $\tau_k(t)$ are bounded time functions, $k = M, S$, $f_k: \mathbb{R}^2 \rightarrow \mathbb{R}^2$ are a smooth maps and $p_1, p_2 \in \mathbb{R}^p$ are parameter sets, which can be time functions. In addition, let us suppose that systems output are given by the smooth functions $y_k = h(x_k)$, which can be linear or nonlinear. B is a vector of suitable length and $u \in \mathbb{R}$ denotes the control command.

Synchronization problem is to lead system (3.26) trajectories to system (3.25) attractor. Now, let us define the synchronization error as follows: $x_i(t) := x_{M,i}(t) - x_{S,i}(t)$. Then the following dynamical system is obtained: $\dot{\boldsymbol{x}} = \Delta f(x, x_k; \pi) + \Delta\tau(t)$ whose output is given by a smooth function $y = h(x)$ and Δ denotes differences between master and slave. Thus, synchronization problem can be addressed by stabilization of the synchronization error system at origin by means of controller u [19]. In this section we study the complete synchronization according to Definition 3.16.

Definition 3.24. All driven second-order chaotic synchronization error systems obtained from (3.25) and (3.26) can be written as

$$\begin{pmatrix} \dot{x}_1 \\ \dot{x}_2 \end{pmatrix} = \begin{pmatrix} x_2 \\ \Delta f(x) + \Theta(x, x_M) \end{pmatrix} + \begin{pmatrix} b_1 \\ b_2 \end{pmatrix} u + \begin{pmatrix} 0 \\ \Delta\tau \end{pmatrix} \tag{3.29}$$

where $\Delta f(x)$ is a smooth and bounded function of the synchronization error, $\Delta\tau$ is the difference between external drive force of the master and slave system and $\Theta(x,x_M)$ results from the discrepancy due to differences of the nonlinear terms.

Remark 3.25. Definition 3.24 includes diverse synchronization schemes. For example: if the dynamical system of the synchronization error $\dot{\boldsymbol{x}} = \Delta f(x,x_k;\pi) + \Delta\tau(t)$ is constructed under assumption of the exact knowledge of the response model (which cannot be restrictive for any application, see [45] and references therein), then $\Delta f(x)$ can be obtained similar than response model. That is, departing from definition $x := x_M - x_S$, one has that $x_S = x - x_M$, then by substitution into the salve function $f_S(x_S)$, Thus, one has that $\Delta f(x,x_k;\pi) = f_M(x_M) - f_S(x-x_M)$. Now, since $f_S(x_S)$ and $f_M(x_M)$ are nonlinear, by defining $\Delta f(x) := f_S(x_S)$, $\Theta(x,x_M)$ results from the deficiency of the nonlinear function. Of course, under assumption of the exact knowledge of the slave function, $\Theta(x,x_M)$ becomes unknown while $\Delta f(x)$ is known.

Theorem 3.26. Let $B = [1\ 0]^T$ be a vector which defines the control channel and $y = x_1 := x_{1,M} - x_{1,S}$. In addition, let us define $e := y - r$ as the stabilization error. The dynamical system of the synchronization error $\dot{\boldsymbol{x}} = \Delta f(x,x_k;\pi) + \Delta\tau(t)$ is locally stabilized at the origin, i.e., $r = 0$ for all $t \geq 0$, under the following dynamical feedback

$$\begin{aligned} u &= ky + k'e \\ \dot{e} &= y - r \end{aligned} \tag{3.30}$$

where the output is $y = x_1$ and k, k' > 0 are control gains.

Proof. From the synchronization error system and the dynamical feedback (3.27), we have

$$\begin{aligned} \dot{x}_1 &= x_2 - kx_1 - k'e \\ \dot{x}_2 &= \Delta f(x) + \Theta(x_M, x) + \Delta\tau(t) \\ \dot{e} &= x_1 - r \end{aligned} \tag{3.31}$$

where $x \in \Omega \subset \mathbb{R}^2$ is a vector whose components are the synchronization error $x_i := x_{i,M} - x_{i,S}$ with $i = 1,2$. System (3.29) can be linearized around the origin for all $t \geq t_0 \geq 0$ to get $z = J(k)z + T$, where $z := z(x,e)$, $T = [0\ \Delta\tau]^T$ and $J(k)$ is the Jacobian and becomes

$$J(K) = \begin{bmatrix} -k & 1 & -k' \\ \dfrac{\partial(\Delta f + \Theta)}{\partial x_1} & \dfrac{\partial(\Delta f + \Theta)}{\partial x_2} & \dfrac{\partial(\Delta f + \Theta)}{\partial e} \\ 1 & 0 & 0 \end{bmatrix}$$

The characteristic polynomial is given by $\lambda^3 + (k - J_{2,1})\lambda^2 + [k(1 - J_{2,2}) - J_{2,1}]\lambda - J_{2,3} - J_{2,2}k = 0$ and without loss of generality we take k' = k. Since the functions $\Delta f(x)$ and $\Theta(x,x_M)$ are contained into an attraction region then they are bounded. The polynomial is perturbed by the terms $J_{i,j}$ for i,j = 1,2,3. Then from perturbation theory, always it is possible to find a value k such that the roots of the polynomial lie in the open-left half complex plane. Now, integrating system (3.29) one has that solution becomes $z(t) = z(0)\exp(-J(k)t) + \int\Delta\tau dt$, where $z(0) = (x(0), e(0))$ denotes initial conditions. Using the triangle inequality one has $\| z(t) \| \leq \| z(0)\exp(-J(k)t) \| + \| \int\Delta\tau dt \|$, where $\|\cdot\|$ denotes the Euclidean norm. Since J(k) is stable, i.e., all its eigenvalues are located in the left-half complex plane, and $\Delta\tau$ is, by definition bounded, hence $\| z(t) \| \leq \| z(0) \| \beta + \| \int\Delta\tau dt \|$. Then the trajectories of system (3.29) are leaded to origin for all $t > 0$, that is synchronization error system is locally stable at origin under controller (3.28) and practical complete synchronization is achieved. □

Remark 3.27. The feedback (3.28) has Proportional-Integral structure. Previous results show that this kind of controller is able to stabilize a chaotic system around origin [21]. In this sense, controller (3.28) yields practical complete synchronization; i.e., for any initial conditions, $x_M(0) \neq x_S(0)$ such that the difference $x_M(0) - x_S(0)$ belongs to a neighborhood $\Omega \subset \mathbb{R}^2$, thus the response system trajectories are close to the drive system trajectories, $x_S(t) \approx x_M(t)$ for all $t \geq t_0 \geq 0$ (where t_0 is the time when controller is activated). Indeed, controller (3.28) leads synchronization error trajectories around origin under following assumptions: (i) Master system order is equal to Slave system order, (ii) Master-Salve parameter mismatches, i.e., dynamical error system is uncertain, (iii) Closed-loop system is single-input-single-output (SISO) and (iv) Synchronization error trajectories are bounded. However, *stability does not mean robustness*. This is, in order to state robust stability properties the closed-loop response should be studied.

Remark 3.28. Consider the following nominal polynomial $P(\lambda)_{\Delta f} = \lambda^3 + (k - \partial\Delta f/\partial x_1)\lambda^2 + [k(1 - \partial\Delta f/\partial x_2) - \partial\Delta f/\partial x_1]\lambda - k\partial\Delta f/\partial x_2 - \partial\Delta f/\partial x_3 = 0$. A sufficient condition for stability is that all coefficients be positive. According to perturbation theory, the characteristic polynomial of J(k) remains stable for a sufficiently small $\partial_i\theta = \partial\theta/\partial x_i$ for $i = 1,2,3$. That is, the polynomial $P(\lambda)_{\Delta f+\Theta} = \lambda^3 + (k - \partial\Delta f/\partial x_1)\lambda^2 + [k(1 - \partial\Delta f/\partial x_2) - \partial(\Delta f + \Theta)/\partial x_1]\lambda - k\partial(\Delta f + \Theta)/\partial x_2 - \partial(\Delta f + \Theta)/\partial x_3 = 0$ is also stable. It should be pointed out that stability result in Theorem 3.26 is local. However, since synchronization problem is interpreted as the stabilization of synchronization error around origin, the local stability analysis makes sense. This fact allows the establishment of the transversal stability via singular values. Therefore, robustness issues are related with the size of $\partial_i\theta = \partial\theta/\partial x_i$. In this sense, the above nominal polynomial $P(\lambda)_{\Delta f}$ results from an *ideal* synchronization error system (i.e., if drive and response system are exactly equal), which is stable at origin for any initial condition $(x(0),e(0)) \in \mathbb{R}^3$ (where $x(0)$ denotes the initial synchronization error and $e(0)$ is the initial stabilization control error) and all $t \geq 0$. In what follows, we will discuss robustness issues departing from property in Theorem 3.26.

Definition 3.29. A system is said to be *well-posed* if its closed-loop transfer function M(s) = C(s)P(s)/(1 + C(s)P(s)) is well defined and proper. That is, the system is stable (i.e., 1 + C(s)P(s) = 0 is Hurwitz) and the numerator degree is less than denominator one (where C(s) = u(s)/e(s) is the transfer function of the controller, P(s) = y(s)/u(s) $\in \Pi$ is the uncertain transfer function of the system, $y(s) = x_{1,M}(s) - x_{1,S}(s)$ is the system output and u(s) is the synchronization force).

Remark 3.30. The closed-loop system is given by dynamics of the synchronization error and dynamical system (3.29). Closed-loop linearization yields the following system: $\dot{z} = J(k)z + T_e$, where $z = [x,e]^T$, J(k) denotes Jacobian matrix (which only depends on control parameters) and $T_e = [0, \tau(t), 0]^T$ is a vector given by external disturbances. Results in Theorem 3.26 implies that there exists a control gain values k and k' such that the closed-loop system (3.29) is well-posed. Such a property is important because it means that rational function P(s) = y(t)/u(s), s = ωj is well defined and proper, which is related with transversal stability (see below).

3.4.2 Robust Stability Analysis of Synchronization Error

As was discussed above, stability does not imply robustness. Robust stability means that the synchronization trajectories x(t) converge to zero for any initial conditions (x(0),e(0)) even in presence of uncertainties (e.g., master/slave mismatches) or disturbances. Here, the robust synchronization is discussed departing from the Small Gain Theorem (SGT) [17], [44] and result in Theorem 3.26. In this sense the proposed measure of robust synchronization is related with transversal stability and is comparable with results in [46]. SGT provides an acceptable and conservative measure of robust stability. Indeed, SGT allows to compute a set of systems whose elements can be synchronized and provides a measure of the family size of systems such that the synchronization manifold from systems (3.29) is invariant.

Now, consider the diagram in Figure 3.19a, where the block Π represents the set of all uncertain plants P, with $P_o \in P \subset \Pi$ the nominal plant (see Remark 3.28) and K the controller (which stabilizes the plant P_o, see Theorem 3.26). Thus, the problem is to find the size of the family Π for given parameters of the controller (3.28) in such way that every plant P ∈ Π be stable, i.e., the robust stability of the synchronized system.

Definition 3.31. A dynamical system P(s) is said to be *robustly stable* if there exists a controller u(s) = C(s)(y(s) - r(s)) such that it stabilizes every member of a family Π containing the uncertain plants P(s).

Definition 3.32. $\mathfrak{R}\mathcal{H}_\infty$ is a subspace with analytic and bounded functions in the open right-half complex plane. The $\mathcal{H}_\infty$-norm of a plant M(s) is defined as $\| M \|_\infty := \sup_\omega \bar{\sigma}[M(j\omega)]$ where $\bar{\sigma}[M(j\omega)]$ represents the largest singular value and M(jω) is the transfer function with $j = \sqrt{-1}$ and ω denotes frequency.

Theorem 3.33. (Small Gain Theorem [17], [44]). Consider a transfer function M(s). Suppose $M \in \mathfrak{R}\mathcal{H}_\infty$ and let σ > 0 be a positive real number. Then system in Figure 3.19b is well-posed for all $\Delta(s) \in \mathfrak{R}\mathcal{H}_\infty$ with: (i) $\| \Delta \|_\infty \leq 1/\sigma$ if and only if $\| M \|_\infty < \sigma$, and (ii) $\| \Delta \|_\infty < 1/\sigma$ if and only if $\| M \|_\infty \leq \sigma$, where M(s) is the

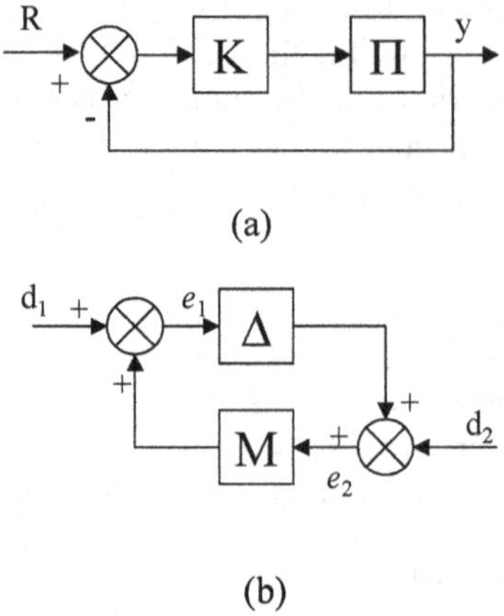

Fig. 3.19. (a) Basic representation of the synchronization problem, (b) representation of the robust synchronization problem

closed-loop system (see Figure 3.19b) and Δ represents the uncertainties of the synchronization system (3.29).

Remark 3.34. In Figure 3.19b, the d_1 and d_2 are the external inputs such as noise and parameter modulation, respectively. That is, the robust stability of the synchronized systems involves synchronization under environmental noise. Moreover, such robust issues include the parameter modulation, which has application in secure communication. In addition, the stabilization error e and the control command u are the inputs to the controller and the plant, respectively. From this point of view, the closed-loop system M(s) represents the synchronization error system under control actions; *i.e.*, system (3.29)), which is stable for any initial conditions $(x(0)\ e(0)) \in \mathbb{R}^3$. Δ can be interpreted as all possible destabilizing uncertainties acting onto synchronization system (3.27), which are represented by $\Theta(x_M,x)$ and/or time-varying parameters, see Remark 3.28.

Proposition 3.35. Assume that the control parameters k and k' are chosen such that the nominal polynomial $\lambda^3 + (k - J_{2,1})\lambda^2 + [k(1 - J_{2,2}) - J_{2,1}]\lambda - J_{2,3} - J_{2,2}k = 0$ is stable. There exists a family of uncertain synchronized systems Π containing $P_o \in P \subset \Pi$ such that the synchronization system (3.29) is well-posed for all Δ(s) with $\| \Delta \|_\infty \leq 1/\sigma$. That is, the stability margin of the synchronized system (3.29) is $1/\sigma$, where σ denotes the maximum singular value for all frequency $\omega > 0$ and any control parameters k and k'.

3.4.3 Illustrative Example

In order to illustrate the robust synchronization analysis by means of SGT and feedback (3.28), we consider two Duffing oscillators. These systems have the same value of the parameters and different initial conditions. In this example we find a measure η of the family Π such that the synchronization system (3.29) is robustly stable. From two Duffing equations, we have the synchronization error system as follows

$$\begin{aligned}
\dot{x}_1 &= x_2 - u \\
\dot{x}_2 &= x_1 - x_1^3 - \delta x_2 + \Theta(x_M, t) \\
\dot{e} &= x_1 - r \\
u &= kx_1 + k'e
\end{aligned} \tag{3.32}$$

where $r = 0$ for all $t \geq 0$, $x_i = x_{i,M} - x_{i,s}$ defines the error between master and slave system and $\Theta(x_M,x,t) = 3x_{1,M}x_1^2 - 3x_{1M}^2 x_1 + \Delta\tau(t)$. The nonlinear term $\Theta(x_M,x,t)$ arises from difference between master and slave systems and lumps the deficiency of the nonlinear function $\Delta f(x)$. System (3.30) is nonlinear and, in order to attain synchronization, the system must be stabilized at the origin. Then, the synchronization system (3.30) can be approached around the origin by

$$\begin{aligned}
\dot{x}_1 &= x_2 - k(x_1 - r) - k'e \\
\dot{x}_2 &= x_1 + \frac{\partial\Theta}{\partial x_1}x_1 - \delta x_2 \\
\dot{e} &= x_1 - r
\end{aligned} \tag{3.33}$$

where $\partial\Theta/\partial x_1 = 6x_{1,M}x_1 - 3x_{1,M}2$, $\partial\Theta/\partial x_2 = 0$, $\partial\Theta/\partial x_3 = 0$. Always it is possible to find some k and k' such that system (3.31) is stabilized at the origin (see Theorem 3.26). That is, $| x_{iM} - x_{iS} | \rightarrow 0$ for all $t \geq 0$. Then the robust stability analysis consists in the determination of the H_∞- norm of the system (3.31). According to Theorem 3.26, one can chose parameters k and k' such that system (3.31) is stable at origin for all $t \geq t_0 \geq 0$. Moreover, singular values $\sigma[M(j\omega)]$ provides the best value for k in order to stabilize a synchronization family.

Figure 3.20 shows the singular values for several k's of system (3.31). Note that for $k = 1$ the destabilizing Δ has a H_∞-norm $\| \Delta \|_\infty = 0.0166$, which means that the largest perturbation has norm less than 0.0166 such that the synchronization system is robustly stable. Nevertheless, for $k = 1$, the size of the synchronization family Π (which is defined by the H_∞-norm of the destabilizing Δ) is considerably small. In Figure 3.20 can be seen that if the control parameter increases then the singular values reduces; consequently the size of the synchronization family Π increases. For instance, for $k = 3$, the H_∞-norm of the uncertainties becomes $\| \Delta \|_\infty = 2/3$. Therefore, the size of the Family Π increases to 3/2, this means that for all nonlinear term $\Theta(x_M,x,t)$ such that $\| \Theta(x_M,x,t) \| < 3/2$, the synchronization system is robustly stable.

Figure 3.21 shows the trajectories of the synchronization system for some initial conditions and some values for the control parameter k. In Figure 3.21a the trajectories are leaded to a neighborhood around origin for $k = 1$. The controller was activated at time $t_0 = 0$ and the initial conditions were chosen as follows: for doted line as $x_{1,M}(0) = 0.03$, $x_{2,M}(0) = 0.045$, $x_{1,S}(0) = 0.05$, $x_{2,S}(0) = -0.015$; for solid line $x_{1,M}(0) = 0.01$, $x_{2,M}(0) = -0.02$, $x_{1,S}(0) = 0.035$, $x_{1,S}(0) = 0.015$; and for dashed line $x_{1,M}(0) = 0.01$, $x_{2,M}(0) = 0.005$, $x_{1,S}(0) = -0.021$, $x_{2,S}(0) = -0.035$. Note that all trajectories lie into the neighborhood of the origin for $t \geq t^* > 0$, where t^* is a small transient time. Such a neighborhood is a ball centred at origin whose radius is $\| \Delta \|_\infty = 0.0166$. This means that, for any initial conditions $(x(0),e(0))$ and $k = 1$, the

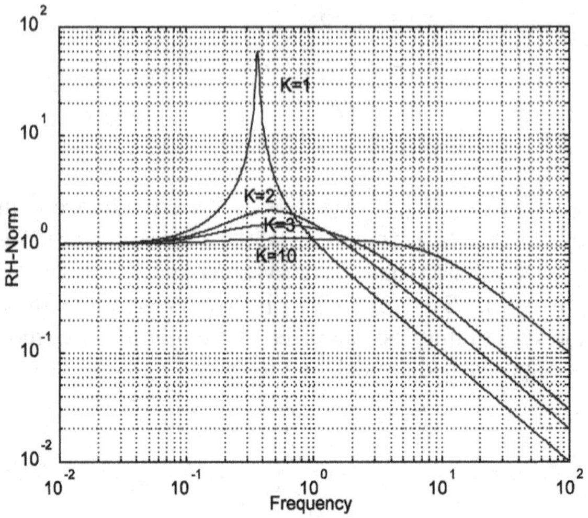

Fig. 3.20. Singular values of the nominal system of the synchronization error around origin for several values of the control parameter K.

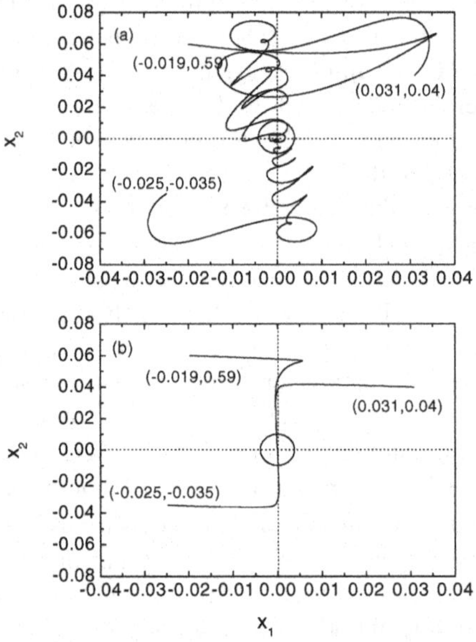

Fig. 3.21. Asymptotic stabilization of the synchronization error trajectories for different initial conditions, (a) k=1 and (b) k=10

synchronization error trajectory does not necessarily reach the origin for $t > t^* > 0$. This fact can be an explanation for some synchronization phenomena [29]. On the other hand, Figure 3.20b shows the trajectories for the same initial conditions than Figure 3.20a and $k = 10$. In this case the trajectories of the synchronization system lie into a ball of radius proportional to $\| \Delta \|_{\infty} = 0.8948$ (see Figure 3.21b). Then from this, the size of the synchronization family Π is greater than the previous case. Therefore trajectories contained into the Π converges to origin. The analysis was carried out departing from tools of the H_{∞} control theory. It was computed a size of a family of synchronization systems, which can be stabilized around the origin by means of a dynamical controller.

A central difference between these results and previous results is the following: *In previous results the goal is to find the "best" parameter set of the control for robust synchronization whereas the present result is a procedure to find the stability margin of the robustness for a given synchronized system.* In addition, it was stated that the size of the family depends on the control parameter. Then the robustness measure is determined by the parameter k. That is, there is similitude with robust synchronization in the Rulkov-Suschik's sense [46]. The SGT provides a measure of robustness of the synchronization scheme against master/slave mismatches. On the other hand, for the case of robustness against parameter variations, the interesting point is, to design a controller which stabilizes every system of the family Π, and every system behaves chaotically.

References

[1] Wu, C.W., Chua, L.O.: A simple way to synchronize chaotic systems with application to secure communication systems. Int. J. of Bifur. and Chaos 3, 1619 (1993)

[2] Solís-Perales, G.: Synchronization in polipode gait, Ms. Sc. Thesis, Universidad Autónoma de San Luis Potosí, México (1999) (in Spanish)

[3] Collins, J.J., Stewart, I.N.: Hexapodal gaits and coupled nonlinear oscillator model. Biol. Cybernetics 68, 287 (1993)

[4] Holstein-Rathlou, N.-H., Yip, K.-P., Sosnovtseva, O.V., Mosekilde, E.: Synchronization phenomena in nephron-nephron interaction. Chaos 11, 417 (2001)

[5] Bragard, J., Boccaletti, S.: Integral behavior for localized synchronization in nonidentical extended systems. Phys. Rev. E 62, 6346 (2000)

[6] Sarasola, C., Torrealdea, F.J., d́Anjou, A., Graña, M.: Cost of synchronizing different chaotic systems. Math. Comp. In Simulation 58, 309 (2002)

[7] Zhou, K., Doyle, J.C., Glover, K.: Robust and optimal control. Prentice-Hall, USA (1996)

[8] Martens, M., Pécou, E., Tresser, C., Workfolk, P.: On the geometry of master-slave synchronization. Chaos 12, 316 (2002)

[9] Josić, K.: Synchronizaiton of chaotic systems and invariant manifolds. Nonlinearity 13, 1321 (2000)

[10] Nijmeijer, H., van der Schaft, A.: Nonlinear dynamical control systems. Springer, USA (1990)

[11] Perez, G., Cerdeirea, H.A.: Extracting messages masked by chaos. Phys. Rev. Lett. 74, 1970 (1995)

[12] Pyragas, K.: Transmission of signals via synchronization of chaotic time delay systems. Int. Jour. of Bifur. and Chaos 8, 1839 (1997)

[13] Rabinovich, M.I., Abarbanel, H.: The role of chaos in neural systems. Neuroscience 87, 5 (1998)

[14] Kapitaniak, T., Skeita, M., Ogorzalek, M.: Monotone synchronization of chaos. Int. Jour. of Bifur. and Chaos 6, 211 (1996)

[15] Mosayebi, F., Qammar, H.K., Hertley, T.T.: Adaptive estimation and synchronization of chaotic systems. Phys. Lett. A 161, 255 (1991)
[16] Cazelles, B., Boudjema, G., Chau, N.P.: Adaptive control of systems in a noisy environment. Phys. Lett. A 196, 326 (1995)
[17] Zhuo, K., Doyle, J.C.: Essential of robust control. Prentice-Hall, USA (1998)
[18] Kocarev, L., Parlitz, U.: General approach for chaotic synchronization with application to communication. Phys. Rev. Lett. 74, 5028 (1995)
[19] Femat, R., Alvarez-Ramirez, J.: Synchronization of a class of strictly different chaotic oscillators. Phys. Lett. A 236, 307 (1997)
[20] Femat, R., Alvarez-Ramírez, J., Fernandez Anaya, G.: Adaptive synchronization of high order chaotic systems: A feedback with low parameterization. Physica D 139, 231 (2000)
[21] Femat, R., Capistrán-Tobías, D., Solís-Perales, G.: Laplace domain controllers for chaos control. Phys. Lett. A 252, 27 (1999)
[22] Koslov, A.K., Shalfeev, V.D., Chua, O.L.: Exact synchronization of mismatched chaotic systems. Int. Jour. of Bifur. and Chaos 6, 569 (1996)
[23] Campos-Delgado, D.U., Femat, R., Martínez-López, F.J.: Laplace domain controllers for chaos control: sub-optimal approaches. IEEE Trans. Circ. and Syst. I (submitted, 2003)
[24] Guckenheimer, J., Holmes, P.: Nonlinear oscillations, dynamical systems and bifurcations of vector fields. Springer, N.Y (1990)
[25] Isidori, A.: Nonlinear Control Systems. Springer, Berlin (1989)
[26] Alvarez-Ramirez, J., Femat, R., Barreiro, A.: A PI controller with disturbance estimation. Ind. Eng. Chem. Res. 36, 3668 (1997)
[27] Haken, H.: Synergetic: an introduction. Springer, Berlin (1983)
[28] Campos-Delgado, D.U., Femat, R., Ruiz-Velazquez, E.: Design of reduced- order controlers via H_∞ and parametric optimisation: Comparison for an active suspension system. In: Landau, I.D., Karimi, A., Hjalmarson, H. (eds.) Special isue on Design and Optmisation of restricted complexity controlers. Eur. J. Control, vol. 9, pp. 48–60 (2003)
[29] Femat, R., Solís-Perales, G.: On the chaos synchronization phenomena. Phys. Let. A 262, 50 (1999)
[30] Martínez-López, F.J.: Suppresion of chaos in third-order dynamical systems and robustnes analysis, Mc. Sc. Thesis, Universidad Autónoma de San Luis Potosí, S.L.P., México (2003) (in Spanish)
[31] Schuster, H.G.: Deterministic chaos, an introduction, Germany (1989)
[32] Brown, R., Chua, L.O.: Clarifying chaos: examples and counterexamples. Int. Jour. of Bifur. and Chaos 6, 219 (1996)
[33] Xiaofeng, G., Lai, C.H.: On the synchronization of different chaotic oscillators. Chaos Solitons and Fractals 11, 1231 (2000)
[34] Femat, R., Alvarez-Ramírez, J., González, J.: A strategy to control chaos in nonlinear driven oscillators with least prior knowledge. Phys. Lett. A 224, 271 (1997)
[35] Pecora, L.M., Carrol, T.L.: Synchronization in Chaotic Systems. Phys. Rev. Lett. 64, 821 (1990)
[36] Rosenblum, M.G., Pikovsky, A.S., Kurths, J.: Synchronization in a population of globaly coupled oscillators. Europhys. Letts. 34, 165 (1996)
[37] van Vreswijk, C.: Partial synchronization in population of pulse-cupled oscillators. Phys. Rev. E 54, 5522 (1996)
[38] Brown, R., Kocarev, L.: A unifying definition of synchronization for dynamical systems. Chaos 10, 344 (2000)
[39] Colonius, F., Kliemann, W.: The Dynamics of Control. Birkhäuser, Basel (2000)

[40] Femat, R.: Chaos in a clas of reacting systems induced by robust asymptotic feedback. Physica D 136, 193 (2000)
[41] Pérez, M., Albertos, P.: Regular and chaotic behavior of a PI-controled CSTR. In: Proc. of the V NOLCOS 2001, St. Petersburg, Rusia, p. 1386 (2001)
[42] Alvarez-Ramírez, J.: Nonlinear feedback for controling the Lorenz equation. Phys. Rev. E 53, 2339 (1994)
[43] Aguirre, L.A., Billings, S.A.: Closed-loop suppresion of chaos in nonlinear driven oscillators. Nonlinear Sci. 5, 189 (1995)
[44] Morari, M., Zafiriou, E.: Robust Proces Control. Prentice-Hall, USA (1989)
[45] Zhou, C., Lai, C.-H.: Extracting messages masked by chaotic signals of time delay systems. Phys. Rev. E 60, 320 (1999)
[46] Rulkov, N.F., Suschik, M.M.: Robustnes of synchronized chaotic systems. Int. Jour. of Bifur. and Chaos 7, 625 (1997)
[47] Abbott, L.F., van Vresweijk, C.: Asynchronous states in networks of pulse- coupled oscillators. Phys. Rev. E 48, 1483 (1993)
[48] Traub, R.D., Wong, R.S.W.: Celular mechanisms of neuronal synchronization in epilepsy. Science 216, 745 (1982)
[49] Short, K.M.: Steps toward unmasking secure communication. Int. J. of Bifur. and Chaos 4, 959 (1994)
[50] Cuomo, K.M., Oppenheim, A.V.: Circuit implementation of synchronization chaos with applications to communications. Phys. Rev. Lett. 71, 65 (1993)
[51] Cuomo, K.M., Oppenheim, A.V., Strogatz, S.H.: Synchronization of Lorenz- based chaotic with application to communications. IEEE Trans. on Circuits and Sistems II 40, 626 (1993)
[52] Zhu, Z., Leung, H.: Adaptive identification of nonlinear systems with application to chaotic communications. IEEE Trans. Circ. and Syst. I 47, 1072 (2000)
[53] Femat, R., Alvarez-Ramírez, J., Castilo Toledo, B., González, J.: On robust chaos suppresion in a clas of nondriven oscillators: Application to Chua's circuit. IEEE Trans. Circuits and Systems I 46, 1150 (1999)
[54] Teel, A., Praly, L.: Tools for semiglobal stabilization by partial state and output feedback. SIAM J. Contr. Opt. 33, 1443 (1995)
[55] Arena, P., Baglio, S., Fortuna, L., Manganaro, G.: Experimental signal transmission using synchronised state controlled celular neural networks. Electronic Lett. 32, 362 (1996)
[56] Torres, L.A.B., Aguirre, L.A.: Inductorles Chuaś circuit. Electronic Letts. 36, 1915 (2000)
[57] Chua, L.O., Yang, T., Zhoung, G.-Q., Wu, C.W.: Adaptive synchronization of Chuaś oscillators. Int. J. of Bifur. and Chaos 6, 189 (1996)
[58] Nijmeijer, H., Mareels, I.M.Y.: An observer looks at synchronization. IEEE Trans. on Circuits and Systems I 44, 882 (1997)
[59] Kocarev, L., Halle, K.S., Eckert, K., Chua, L.O., Parlitz, U.: Experimental demostration of secure communication via chaotic synchronization. Int. J. of Bifur. and Chaos 2, 709 (1992)
[60] Suykens, J.A., Curan, P.F., Vandewalle, J., Chua, L.O.: Nonlinear H_∞ synchronization of chaotic Luré systems. Int. Jour. of Bifur. and Chaos 7, 1323 (1997)
[61] Suykens, J.A.K., Curran, P.F., Chua, L.O.: Robust nonlinear H_∞ synchronization of the chaotic Luré systems. Trans. on Circuits and Systems I 44, 891 (1997)
[62] van der Schaft, A.: L_2-gain and pasivity techniques in nonlinear control, 2nd edn. Springer, London (2000)

4 Robust Synchronization Via Geometrical Control: A General Framework

4.1 Synchronization of Second-Order Driven Systems with Different Model

Particular interest has been devoted to study chaos synchronization of similar oscillators (see for instance [1], [2], [3], [4] and references therein). Such synchronization strategies have potential applications in several areas such as secure communication [5], [6], [7] biological oscillators and animal gait [8], [9]. It has been shown that two identical chaotic oscillators can be synchronized [10]. The general framework is developed toward synchronization of chaotic systems with different model. In seek of clarity and completeness in presentation, the second-order driven oscillators are considered for synchronization feedback.

Although, in most cases the exact values of parameters are unknown, some strategies have been reported to compensate mismatches in parameters [11], [12], [13]. Several authors have reported adaptive estimation techniques to attain the chaos synchronization when the model parameters are unknown [11] or they are time-varying [12]. Nevertheless, the adaptive strategies have the following drawback: *The parameters structure must be known*. In others, it can be desirable a control scheme which allows the synchronization in spite of chaotic oscillators have not the same structure. As a consequence, their models are not similar (for example, the multimode laser, oscillatory neural systems, or chemical processes) [13]. On the other hand, as we discuss in previous chapters, a strategy based on robust asymptotic stabilization has been used to suppress and synchronize chaos. The control scheme comprises a linearizing control law and an uncertainty estimator. The main idea to deal with the uncertainties, is to lump them in a nonlinear function in such way that it can be interpreted as a *'state'* in an externally dynamically equivalent system. Then, an estimate of the uncertainties is obtained (via the *new state*) by means of a state estimator. An advantage of this robust strategy is the following: *The controller requires least prior information about the system model* [14]. Besides, it has been proved that the robust asymptotic control can be used to suppress chaos in high order chaotic systems [15].

R. Femat & G. Solis-Perales: Robust Syn. of Chaotic Sys. Via Feedback, LNCIS 378, pp. 99–137, 2008.
springerlink.com

The aim of this section is to show that the synchronization of a class of *strictly different* chaotic oscillators can be performed. That is, two chaotic systems of the same order which have different model. In particular, we consider that the master model is represented by the Duffing equation

$$\ddot{x}_D + \delta_D \dot{x}_D - x_D + x_D^3 = \gamma_D \cos(\omega_D t) \tag{4.1}$$

and the slave oscillator is given by the van der Pol equation

$$\ddot{x}_V + \delta \dot{x}_V (1 - x_V)\dot{x}_V + x_V^3 = \gamma_V \cos(\omega_V t) + u \tag{4.2}$$

Both, Duffing and van der Pol oscillators were synchronized in chapter 3. However, such a design procedure is restricted to Assumption S.3 (see pag. 58), which implies that control is acting onto the same channel than the measure state. Here, Assumption S.3 is relaxed in order to develop a more general design algorithm for synchronization.

4.1.1 Problem Statement

Here, we use a control strategy to attain synchronization of the oscillators (4.1) and (4.2). The synchronization strategy is based on input-output linearization design techniques [16], [17]. The problem of chaos synchronization is seen as a stabilization one where the control objective is to lead the trajectories of the synchronization system (i.e., the difference between the Duffing and van der Pol equations) at the origin. To this end, let us define $(x_{1,M}, x_{2,M}) = (x_D, x_D)$ and $(x_{1,S}, x_{2,S}) = (x_V, x_V)$. In this way, both Duffing and van der Pol equations can be rewritten in the following form

$$\begin{aligned} \dot{x}_1 &= x_2 \\ \dot{x}_2 &= f(x_1, x_2; \pi_1) + \tau(t; \pi_2) - u \end{aligned} \tag{4.3}$$

where $f(x_1,x_2;\pi_1)$ is a nonlinear smooth function, $\tau(t;\pi_2)$ is an external excitation force, and π_1 and π_2 are parameter sets. Of course, depending on the values of the parameters, the system (4.3) can display chaotic behavior.

Remark 4.1. Note that system (4.3) can be written as $\dot{x} = F(x;\pi) + Bu + T(t)$ where $F(x;\pi) = (x_2\ f(x;\pi))^T$ and $g(x) = (0\ 1)^T$ are smooth for any $x \in \mathbb{R}^2$. In addition, $T(t) = (0\ \tau(t))^T$ can be interpreted as a disturbance vector, which is continuos for all time $t \geq 0$.

Lemma 4.2. Let $y = x_1$ be the output of the synchronization error system. There is a coordinate transformation given by the map $\Phi : \mathbb{R}^2 \to \mathbb{R}^2$ which is defined as $z = \Phi(x)$, where $x = \Phi^{-1}(\Phi(x))$ for any x and z belonging to $\mathbb{R}^2$ such that $\dot{z}_1 = z_2$, $\dot{z}_2 = L_F^2 h(x) + L_F L_g h(x) u$ where $h(x) = y$ and $L_F h(x) = (\partial h/\partial x)F$ denotes the Lie derivative of the output $h(x)$ along the vector field $g(x)$ and $L_g L_F h(x)$ is the Lie derivative of the nonlinear function $L_g h(x)$ along F.

Proof. Let us define $h(x) := y = x_1$ as the output of the synchronization error system. The first Lie derivative of the system output y results $y = \partial h(x)/\partial x$, $x = \partial h(x)/\partial x$ $[F(x;\pi) + g(x)u + T(t)] = \partial h(x)/\partial x\ F(x) + \partial h(x)/\partial x\ g(x)u + \partial h(x)/\partial x$, $T(t) = L_F h(x) + L_g h(x) + L_T h(x)$. Since $L_g h(x) \equiv 0$ for all $x \in \mathbb{R}^2$, $y = L_F h(x) = x_2$. By defining $\phi_1(x) = L0F h(x) = h(x)$ and $\phi_2(x) = L_F h(x)$, one has that $\Phi(x) = (\phi_1(x)\ \phi_2(x))^T$ defines the coordinates transformation z:= $\Phi(x)$, which is a diffeomorfic map $\Phi(x)$: $\mathbb{R}^2 \rightarrow \mathbb{R}^2$ for all $x \in \mathbb{R}^2$. Now, taking the second derivative of the system output, y = $\partial L_F h(x)/\partial x$ x = $\partial L_F h(x)/\partial x$ [F(x;π) + g(x)u] = $L_F^2 h(x)$ + $L_g L_F h(x)u$. Now, since z = $\Phi(x)$ the synchronization error system can be written in transformed coordinates as follows

$$\dot{z}_1 = z_2$$
$$\dot{z}_2 = L_F^2 h(x) + L_g L_F h(x)u \tag{4.4}$$

Remark 4.3. Note that for the second-order driven systems in the form (4.1) minus (4.2), one has that L2F$h(x)$ =$f_M(x)$ - $f_S(x)$ + $\gamma_D\cos(\omega_D t)$ - $\gamma_v\cos(\omega_v t)$ and $L_g L_F^2 h(x) \equiv 1$ for all $x \in \mathbb{R}^2$ and $t \geq 0$.

The synchronization problem is to find a feedback control such that z →0 as t → ∞. The theory of nonlinear feedback control is fully develop. In particular state feedback requires the measurements and feedback of the state vector via control law. The control law counteracts the nonlinearities and induces a desired behavior. However, if the vector fields are uncertain, nonlinearities can not be exactly compensated. As a consequence the feedback control law could not induce the desired behavior. In our study case, the system of the synchronization error can be stabilized by the following linearizing control law

$$u = -(L_g L_F h(x))^{-1}(L_F^2 h(x) + Kz) \tag{4.5}$$

where the control constants $K \in \mathbb{R}^2$ are chosen in such way that $P_2(s) = s^2 + k_2 s + k_1 = 0$ is a Hurwitz polynomial (that is, $P_2(s)$ has all its roots in the open left-half complex plane). Thus, the closed-loop system can be written as z = Az, where $A \in \mathbb{R}^{2\times 2}$ is the companion matrix of the coefficients $K \in \mathbb{R}^2$ whose characteristic equation is given by the polynomial $P_2(s)$.

The linearizing control law (4.5) has a nice feature. It induces a linear behavior to the nonlinear system (4.4). However, in the most practical cases, the states of the vector $z \in \mathbb{R}^2$ are not available for feedback from measurements. So that, control law (4.5) can not be directly implemented. To avoid this problem, nonlinear state observability theory was developed [17]. A nonlinear system is said *observable* if it satisfies the observability rank condition [18]. Observability is a global concept. It implies that the dynamics of the unmeasured states can be reconstructed from measured states. Some works regarding the stabilization of nonlinear systems via output (measured) feedback have been reported [19]. The idea is to design a high-gain state estimator in order to observe the unmeasured states. In this way, for system (4.4) the following high-gain estimator can be obtained [14][15]

$$\dot{\hat{z}}_1 = \hat{z}_2 + Lk_2(z_1 - \hat{z}_1)$$
$$\dot{\hat{z}}_2 = L_F^2 h(x) + L_g L_F h(x)u + L^2 k_2(z_1 - \hat{z}_1) \tag{4.6}$$

where $\hat{z} = (\hat{z}_1, \hat{z}_2)$ is an estimated of $z = (z_1, z_2)$, $L > 0$ is an adjustable estimation parameter, k_1 and k_2 are estimation constants which are chosen in such way that the polynomial $s^2 + k_1 s + k_2 = 0$ has all its roots in the open left-hand complex plane. Note that the state estimator (4.6) reconstructs asymptotically the dynamics of the state z_2 (which is associated with velocity) from the measurements of z_1 (which means position). Thus the feedback control law becomes

$$u = -(L_g L_F h(\hat{x}))^{-1}(L_F^2 h(\hat{x}) + K\hat{z}) \tag{4.7}$$

where $\hat{x} = \Phi^{-1}(\hat{z})$, $k>0$,, and $\hat{z}$ are estimated values of the actual states $z \in \mathbb{R}^2$. Feedback control based on high-gain observers can induce undesirable dynamics effects such as the so-called *peaking phenomenon* [20]. This phenomenon leads to closed-loop instabilities which are represented by time-finite escapes and large overshooting. To diminish the effect of these instabilities, the controller (4.7) is often modified using the following saturation function: $u^S = \{-(L_g L_F h(\hat{x}))^{-1}(L_F^2 h(\hat{x}) + K\hat{z})\}$, where Sat $\{\bullet\}$: $\mathbb{R} \to B \subset \mathbb{R}$ is a bounded set containing the origin. Moreover, it is possible to prove that the output feedback u^S leads semiglobal asymptotic stabilization of the system (4.4) (for a sketch of the proof see Theorem 4 in [19]).

The high-gain observer (4.6) solves the problem of the unmeasured states $z \in \mathbb{R}^2$ by providing an estimate of them from measurements (position). Nevertheless, the controller (4.7) requires knowledge about the differences between the master and slave oscillators (that is, $L_F^2 h(x)$ and $L_g L_F h(x)$). Hence, the feedback law (4.7) can not be physically implemented because it does not solve the problem associated with model uncertainties.

In what follows, a robust input-output linearization feedback is developed. The proposed robust controller yields asymptotical stability to the closed-loop system and it is based on uncertainties estimation. First, the uncertain terms $L_F^2 h(x)$ and $L_g L_F h(x)$ are lumped in a nonlinear function. Then, the lumping nonlinear function is interpreted as a variable state in an extended system (3-dimensional) which is externally dynamically equivalent to system (4.4). In chapter 2 it has been shown that this procedure results in chaos suppression, then the stabilization of synchronization error system around origin is expected. Hence, synchronization can be achieved.

4.1.2 Synchronization Despite Unknown Master Model

Since the synchronization problem is saw as stabilization one, in what follows, we state a robust asymptotic stabilization technique (RAS) with least prior knowledge. It has been proved that under the RAS, a chaotic system can be practically stabilized [15]. This means that states of the synchronization-error system (4.4) approach to a bounded region around the origin, (z_1 and z_2 tend asymptotically towards a closed region around zero). The proposed synchronization strategy is based on the input-output control theory presented in Chapter 2.

Let us assume the following:

A.1) The output (measurement) of the synchronization error system is $y = z_1 = x_1$

A.2) The master model is unknown then the field $F(x;\pi)$ is uncertain. Hence, the Lie derivative of this vector field is also uncertain.

The assumption (*A.1*) is realistic because in most cases only the position is available for feedback from the master as well as slave oscillator. Although time-derivative of position can be obtained by means of encoder, the procedure is very sensitive to noisy measurements. Concerning the assumptions (*A.2*), it is clear that the slave system does not know the master model. Hence, the differences between master and slave are obviously uncertain, and the feedback controller neither exactly counteract them nor induce a linear behavior.

The idea to deal with the uncertain terms, is to lump them in a new nonlinear function as follows

$$\begin{aligned}\dot{z}_1 &= z_2\\ \dot{z}_2 &= \Theta(z,t;P) - u\end{aligned} \tag{4.8}$$

where $\Theta(z,t;P) = L_F^2 h(x) + L_g L_F h(x)$. Defining $\eta := \Theta(z,t;P)$, the lumping term can be interpreted as a new state in such way that the system (4.8) can be rewritten as the following (externally dynamically equivalent) system

$$\begin{aligned}\dot{z}_1 &= z_2\\ \dot{z}_2 &= \eta - u\\ \dot{\eta} &= \Xi(z,\eta,u,t;P)\end{aligned} \tag{4.9}$$

where $\Xi(z,\eta,u,t;P) = z_2\, \partial_1\Theta(z,t;P) + [\eta - u]\partial_2\Theta(z,t;P)$, $\partial_k\Theta(z,t;P) = \partial\Theta(z,t;P)/\partial z_k$, $k = 1,2$. It is straightforward to prove that $\Psi(z,\eta,t;P) = \eta - \Theta(z,t;P)$ is a first-integral of the system (4.9). This means that along the trajectories of system (4.9), one has that $d\Psi(z,\eta,t;P)/dt = 0$ for all $t \geq 0$. Hence, the states representation (4.9) is externally dynamically equivalent to system (4.9) (and consequently to system (4.4)), as long as initial conditions satisfy $\Psi(z_0,\eta_0,0;P) = 0$. This implies that the dynamics of the new state η provides the dynamics of the lumping function $\Theta(z,t;P)$ which involves the uncertain terms $L_F^2 h(x) + L_g L_F h(x)$.

Remark 4.4. System (4.9) has the following properties: (a) A geometrical interpretation of the state representation (4.9) can be given. If z(t) is a solution of the system (4.8) with initial conditions $z_0 = z(0)$ and $(z(t),\eta(t))$ is a solutions of the system (4.9) with initial conditions $(z_0,\eta_0) = (z(0),\eta(0))$, then the equality $\Psi(z_0,\eta_0,0;P) = 0$ and the condition $d\Psi(z,\eta,t;P)/dt = 0$ imply that the solution z(t) is a projection of the solution $(z(t),\eta(t))$. (b) The uncertainties have been lumped in an uncertain function which can be estimated by means of an unmeasured but *observable* state. Furthermore, if one would be able to stabilize system (4.9) without making use of the algebraic constraint $\Psi(z,\eta,t;P)$, one would be able to stabilize system (4.8) and its equivalent system (4.4).

Now, let us consider the following linearizing control law

$$u = \eta + \boldsymbol{K}^T z \tag{4.10}$$

where the control constants $K \in \mathbb{R}^2$ are chosen as in the eq. (4.5). If initial conditions satisfy $\Psi(z_0,\eta_0,0;P) = 0$, then under the feedback control law (4.10), (z,η) converges exponentially to zero. Convergence of $\eta(t)$ to zero follows from the fact that the closed-loop system is in cascade form [15]. The control dynamics is given by $u = \Xi(z,\eta,u,t;P) + K^T z$. Since $\Theta(z,t;P)$ is a smooth function which is contained into the attractor, hence $\Theta(z,t;P)$ and its derivatives $\partial\Theta(z,t;P)/\partial z_k$, $k = 1,2$, are bounded. This means that $\Xi(z,\eta,u,t;P)$ is a smooth and bounded function, hence u is also a smooth function and bounded. Consequently, $\Psi(z,\eta,t;P) = \eta - \Theta(z,t;P) = 0$ is a first-integral of the system (4.9). In addition, since $\eta = \Theta(z,t;P)$, hence η is bounded. Finally, since z(t) converges exponentially to zero, $\eta(t)$ also converges exponentially to zero. Nevertheless, once again the linearizing feedback control law (4.10) is not realizable because it requires measurements of the states (z_1,z_2,η). However, as it was established in equation (4.6), the problem of estimating unmeasured states can be addressed by using a high-gain observer. The dynamics of the states z_2 and η can be reconstructed from the output $y = z_1$. The state estimator is written in the following form

$$\begin{aligned}\dot{\hat{z}}_1 &= \hat{z}_2 + L\kappa_1(z_1 - \hat{z}_1)\\ \dot{\hat{z}}_2 &= \hat{\eta} - u + L^2\kappa_2(z_1 - \hat{z}_1)\\ \dot{\hat{\eta}} &= L^3\kappa_3(z_1 - \hat{z}_1)\end{aligned} \tag{4.11}$$

and the linearizing feedback control law becomes

$$u = \hat{\eta} + \boldsymbol{K}^T\hat{z} \tag{4.12}$$

The above control law can be modified by the saturation function $\mathrm{Sat}\{\hat{\eta} + K^T\hat{z}\}$ to diminish the effects of the peaking phenomenon [20].

Notice that the linearizing control law (4.12) only uses estimated values of the uncertain terms $L_F^2h(x) + L_gL_Fh(x)$ (by means of $\hat{\eta}$) and $\hat{z}$. Thus, the RAS is given by the dynamic compensator (4.11) and the linearizing control law (4.12). The controller (4.11),(4.12) yields practical stabilization, i.e., the states of the system (4.90) converges to a ball B whose radius is of the order on L^{-1}. Additionally, note that it is not necessary to construct the uncertain system (4.9). In fact, the extended system is used to prove closed-loop stability and analyze the properties of the RAS.

Proposition 4.5. Let us define $e_1 = Q^2(z_1 - \hat{z}_1)$, $e_2 = Q(z_2 - \hat{z}_2)$ and $e_3 = \eta - \hat{\eta}$. Then, the closed-loop system is represented by $\dot{e} = QA(\kappa)e + \Phi(z,\eta,u,t;P)$, where $\Phi(z,\eta,u,t;P) = [0,0,\Xi(z,\eta,u,t;P)]^T$ is bounded, $e = [e_1,e_2,e_3]^T$ and the companion matrix $A(\kappa)$ is given by

$$A(\kappa) = \begin{bmatrix} -\kappa_1 & 1 & 0\\ -\kappa_2 & 0 & 1\\ -\kappa_3 & 0 & 0\end{bmatrix} \tag{4.13}$$

where the estimation constants κ_i, i = 1,2,3, are chosen in such way that the polynomial $s^3 + \kappa_3 s^2 + \kappa_2 s + \kappa_1 = 0$ has all its roots located at the open left-hand complex plane. Since, z belongs to some attractor, then $L_F^2h(x)$ and $L_gL_Fh(x)$ are bounded functions. Hence, $\Theta(z,t;P)$ and $\Xi(z,\eta,u,t;P)$ also are bounded functions. Additionally, since $A(\kappa)$ has all its eigenvalues at the left-half complex plane, we can conclude that $\lim_{t\to\infty} e \to 0$. That is, the estimation error e is globally asymptotically stable (GAS) at zero which implies that $\hat{z} \to z$ and $\eta \to \hat{\eta}$, in consequence, the feedback control (4.12) tends to the linearizing controller given by equation (4.7). Then, control actions counteract the nonlinear uncertainties and induce a linear behavior.

Finally, note that since $\Theta(z,t;P)$ is uncertain, the function $\Xi(z,\eta,u,t;P)$ is correspondingly unknown. Thus, such term has been neglected in the construction of the observer (4.11). As the case of output feedback control, one can use Theorem 4 in [19] to conclude that the saturated version of the controller (4.12) yields semiglobal asymptotic practical stabilization of the uncertain system (4.4).

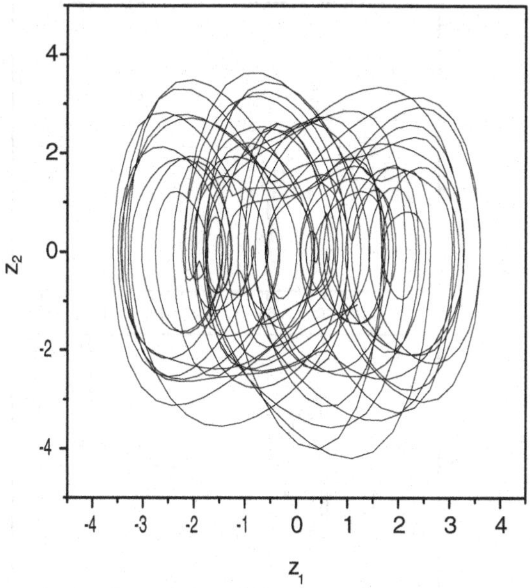

Fig. 4.1. Phase portrait of the synchronization error system

4.1.3 Numerical Simulations

Figure 4.1 shows the phase portrait of the synchronization error system. In order to achieve synchronization, the trajectories in Figure 4.1 must be stabilized at origin. Figure 4.2 shows the stabilization of the synchronization error system (4.4).

The parameters of the master and slave oscillators were taken as in the simulation to construct the phase portrait in Figure 4.1. The control gains were chosen as $k_1 = 6.0$ and $k_2 = 8.0$, such that the polynomial $P_2(s)$ be Hurwitz. The values of the estimator parameters were chosen as: $\kappa_1 = 3.0$, $\kappa_2 = 3.0$ and $\kappa_3 = 1.0$. Then the eigenvalues of the companion matrix (4.13) are located at -1.0. The value of the high-gain estimation parameter $L = 35$. The control was activated at $t \geq 60$. Note that the chaotic dynamics of the synchronization error is suppressed. Besides, the states z_1 and z_2 are contained in a closed region around zero (practical stability). The oscillations after the controller is activated are due to the states of the system (4.9) converges to a ball *B* whose radius is proportional to L^{-1}. This means that: *since the states of the system (4.9) converges to a ball whose radius is proportional to* O^{-1}*, then the states of the system (4.4) converges to the same ball. Of course, if O increases the radius of the ball decreases and, consequently, the trajectories of the states* (z_1,z_2) *are closer to origin.*

The synchronization command, u, is presented in the Figure 4.3. Note that the controller "must pay the cost" of the differences between the master and slave models. Since the controller counteracts the differences between the master and slave oscillators, it absorbs the energy "excess" from the slave system. To illustrate the real chaos synchronization, we have plotted the position and velocity of both master and slave oscillators (see Figure 4.4). Note that the van der Pol trajectory tracks the trajectory of the Duffing system (the master).

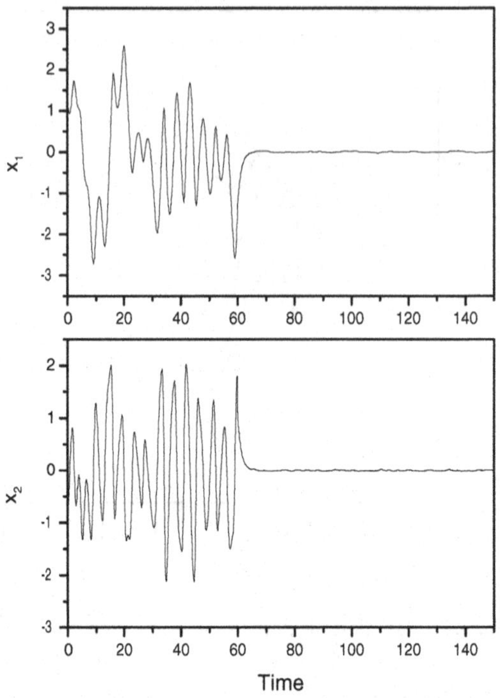

Fig. 4.2. Chaos stabilization of the synchronization error system, the unstable orbits have been stabilized around origin (Practical synchronization of van der Pol and Duffing equation)

4.2 Adaptive Synchronization of High-Order Chaotic Systems: A Feedback with Low Order Parameterization

A general procedure has been proposed for synchronizing second-order driven oscillators. In what follows a general framework is developed to synchronize high-order chaotic systems. Due to synchronization has potential interest in several areas such as secure communication [6], [7], biological oscillators [21] and animal gaits [9] and [10]. Recently, synchronization of chaotic systems and its application to secure communication has received much attention. Since 1990 when master-slave synchronization was introduced by Pecora and Carroll [1] transmission of signals via a chaotic carrier has been widely studied. The main goal has been to mask the message signal via the transmitter which is a chaotic system [22]. The security level of the transmitter/receiver scheme can accommodate message signals given by triangle wave, period doubled signal, and digital signal which means that n-periodic signals can be masked by chaotic carrier. Moreover chaotic and voice signals can be transmitted via chaos-based schemes. Chaos-based secure communication has been tested via the power spectrum with promising results [23]. The actual research can be focussed in two main

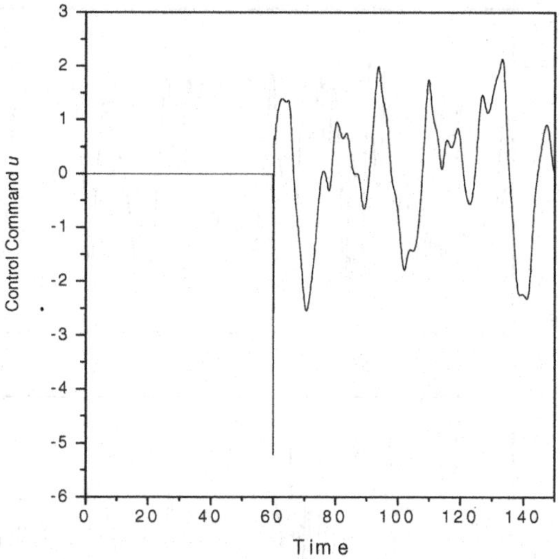

Fig. 4.3. Control command for chaos synchronization between van der Pol and Duffing oscillator

objectives: a) Development of synchronization systems and b) Understanding the synchronization phenomenon. In both development and understanding cases, the synchronization problem is a nonlinear control one. From the control theory point of view, the synchronization problem can be stated as a *tracking* problem or as a *stabilization* problem. This is, a control command yields the synchronization.

It has been shown that two identical chaotic oscillators can be synchronized. Monotone synchronization is a feedback scheme based on high-gain control [3]. Monotone synchronization has application to secure communication; however, it is based on high-gain feedback which is very sensitive to noise. Nevertheless, most of the dynamical systems have model (or parametric) uncertainties. Then, one can expect that synchronization systems are represented by non-identical models. To avoid this problem, some strategies have been reported. In particular, several authors have reported adaptive estimation techniques (for example, [5], [11], [12]). These techniques present an acceptable performance and allow synchronization despite the parameters are not known [3] or they are time-varying [11]. Nevertheless, such strategies have a drawback: *the parameters structure on the model should be known.* This requirement results in very complex feedback schemes. Indeed, feedback control from adaptive algorithms can be quiet complex even for non-chaotic nonlinear systems. For example, for nonlinear systems with nonlinear parameter structure, projections algorithms are required for stabilizing around a prescribed point tracking a desired trajectory.

Although the parameters structure can be known in some cases, it would be desirable to have a feedback scheme in order to achieve synchronization in spite of the slave oscillator has least prior knowledge about the structure of the master system. This necessity of robustness can be required in some systems (for instance, the multimode laser, animal gait or oscillatory neural systems). Moreover, in many real systems, the synchronization is carried out in spite of the oscillators are strictly

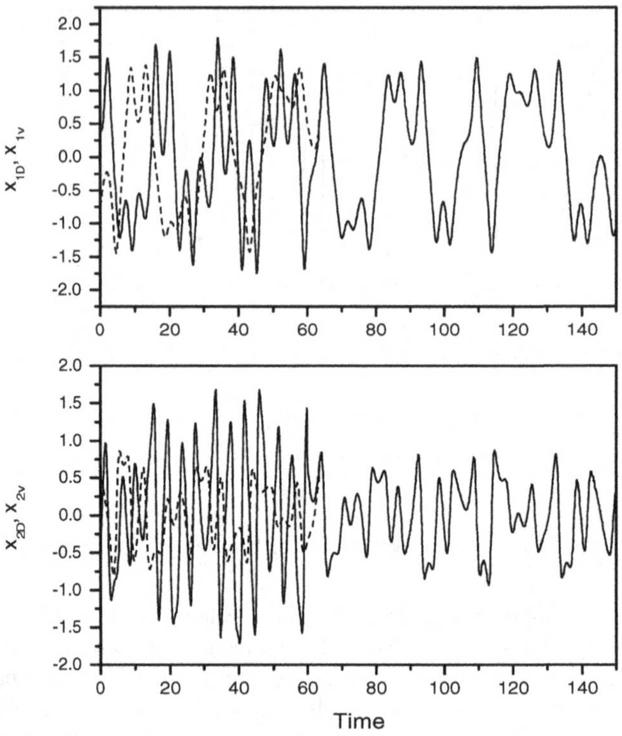

Fig. 4.4. Time evolution of the synchronization behavior, dashed line represents the slave system (van der Pol) and continuous line the master system (Duffing system)

different. For example, biological oscillators are often synchronous even if the master and slave systems are quite different. Synchronization is attained but the slave system does not known the structure of the master oscillator (see [21] for the case of neuron oscillators or [24]).

On the other hand, a robust asymptotic continuous feedback (RAF) has been recently developed. Such controller allows the chaos suppression of second-order driven oscillator (see Sect. 1). Moreover, the synchronization can be achieved even if both master and slave systems are strictly different. The RAF scheme does not require a priori information about the system model to carry out the chaos suppression neither knowledge of the parameter structure nor more of one measured state. Hence, it is a good choice to perform chaos synchronization. In this sense, the RAF can be used to study synchronization between unknown systems (for instance, the synchronization between neuron clusters, [21], or β - cells in pancreas [25]).

Following the main idea reported in previous sections, an algorithm is developed for synchronizing chaotic systems of finite dimension. The proposed scheme is based on the lumping of the unknown functions into a nonlinear function in such way that the lumping function can be interpreted as a '*state*' in an extended system. Then, an estimate of the uncertainties is obtained (via the *augmented state*) by means of a high-gain estimator. The proposed scheme has the following advantages: (i) The controller does not require *a priori* information of the unknown functions neither parameter structure, and (ii) Only one controller parameter is required to tune. In this way, like previously reported adaptive schemes, the proposed scheme allows the synchronization of the non-identical chaotic systems. However, the proposed adaptive strategy is easier to tune and requires least prior knowledge. In principle, since the proposed adaptive controller is simple. The proposed feedback provides information regarding the chaos control. For instance, the knowledge about the synchronization between neurons [21] could be understood by the unveiling of a transmitted signal (see the decoding procedure studied below), unmasking messages in a chaotic carrier [26] or mechanisms of nephron-nephron synchronization [24].

4.2.1 Preliminaries

Here, we review some results on the feedback stabilization of uncertain nonlinear systems through state/uncertainty estimation. To this end, let us consider the following nonlinear dynamical system whose dynamical behavior is chaotic

$$\begin{aligned} \dot{x}_M &= F_M(x_M;\pi_{1,M}) + T_M(t;\pi_{2,M}) \\ y_M &= C_M x_M \end{aligned} \tag{4.14}$$

where $x_M \in \mathbb{R}^n$ is a state vector, $F_M(x_M;\pi_{1,M})$ is a smooth vector field, $T_M(t;\pi_{1,M})$ is an external exciting force, $\pi_{1,M}$ and $\pi_{2,M}$ are parameter set and $y_M \in \mathbb{R}$ is the output system (measured state). C_M is a vector of proper length which defines the output channel. Note that, without loss of generality, we can assume that the measured state is given by $y_M = x_{1,M}$.

Let us now take a chaotic dynamical system of the same order than (4.14)

$$\begin{aligned} \dot{x} &= \Delta F(x;\pi_1) + \Delta T(t;\pi_2) - B \\ y &= x_1 \end{aligned} \tag{4.15}$$

where $x_S \in \mathbb{R}^n$ denotes the state vector of the slave system, B is a vector of suitable size which defines the control channel, $u \in \mathbb{R}$ is the control command, and the parameters $\pi_{1,M} \neq \pi_{1,S}$ and $\pi_{2,M} \neq \pi_{2,S}$. The vector C_S defines the measured state of the slave system. Note that we can assume that $C_S = C_M$. As in previous chapters, system (4.14) is the so-called "*master*" whereas eq. (4.15) represents the "*slave*" system. In the spirit of Mettin *et al.*, [27], system (4.14) describes the goal dynamics while the system (4.15) represents the experimental system to be controlled. The synchronization problem can be stated as follows: *Given the transmitted signal* y_M *and least prior information about the structure of the nonlinear filter, system (4.14), to design a receiver signal* $u(t)$ *which synchronizes the orbits of both the master and slave systems.*

From the control theory viewpoint, the synchronization problem can be seem as follows: Let us define $x \in \mathbb{R}^n$ such that $x_i = x_{i,M} - x_{i,S}$ for i = 1,2,3,…,n. Then, the following dynamical system describes the dynamics of the synchronization error

$$\dot{x} = \Delta F(x;\pi_1) + \Delta T(t;\pi_2) - B$$
$$y = x_1 \tag{4.16}$$

where $\Delta F(x;\pi_1) = F_M(x_M;\pi_M) - F_S(x_S;\pi_S)$ is a smooth vector field, $\Delta T(t;\pi_2) = T_M(t;\pi_M) - T_S(t,\pi_S)$, which is the difference between the external exciting forces. In this way, the synchronization problem can be seem as the stabilization of the Eq. (4.16) at the origin. In others words, *the problem is to find a feedback control law* $u = u(t)$ *such that* $\lim_{t\to\infty} x \to 0$ *(which implies that* $x_S \to x_M$*) as* $t \to \infty$. The goal is to stabilize at origin, the discrepancy between driving and response system in such way that synchronization is achieved. Synchronization error system is defined as the dynamical difference between drive and response systems and includes: (i) model mismatches, which means that the model of the drive system cannot be the same than response system, (ii) unknown initial conditions, which implies that time series of drive system cannot be equal than response system, (iii) parametric uncertainty, which means that response system could comprise inaccuracy construction. Mettin *et al.*, [28] applied a generalized approach to solve the above posed synchronization problem (which is, actually, a control problem) where parametric dependence on the control forces and no detailed model of the system are considered. A drawback of the Mettin et. al.'s approach is that it leads to *nonfeedback* control strategies, which have not guaranteed stability margins.

The approach developed in this book considers uncomplete state measurement and no detailed model of the system to guarantee robust stability (in fact, robust synchronization). Our approach includes a state/uncertainty observer and leads to a *robust feedback* control scheme. Here, we show that general procedure can result, under suitable conditions in the controller discussed in previous Chapters.

Significant advances have been made in the theory of nonlinear feedback control. The nonlinear state feedback theory is fully developed (see [16], [17] for a comprehensive introduction to nonlinear geometrical control theory). Results in [28], show the application of geometrical control on synchronization and suppression of chaos, respectively. Here, we recall the basic concepts.

Let us consider the following nonlinear dynamical system

$$\dot{x} = F(x;\pi_1) + g(x;\pi_2)u$$
$$y = x_1 \tag{4.17}$$

where $x \in \mathbb{R}^n$, $u \in \mathbb{R}$, $y \in \mathbb{R}$ is the system output and π_1, π_2 are parameter sets. In addition, following ideas in sect 2.4 the invertible coordinates transformation

$$z = \Phi(x;\pi) \tag{4.18}$$

can be found in such manner that system (4.17) can be written in the following canonical form [17]

$$\begin{aligned} \dot{z}_i &= z_{i+1}; \qquad 1 < i \le \rho - 1 \\ \dot{z}_\rho &= \alpha(z,v;\pi_3) + \beta(z,v;\pi_4)u \\ \dot{v} &= \varsigma(z,v) \\ y &= z_1 \end{aligned} \tag{4.19}$$

where $y \in \mathbb{R}$ is the system output, ρ is the relative degree of the system (4.17) (ρ is equal to the lowest order time-derivative of the output y that is directly related to the control u), α and β are in general nonlinear functions with β nonsingular for all $\pi \in \mathbb{R}^q$, π is a parameter set, $v \in \mathbb{R}^{n-\rho}$ is the unobservable states vector (internal dynamics).

Then, one can design the following feedback controller

$$u = -\frac{1}{\beta(z,v;\pi_4)}\left[\alpha(z,v;\pi_3) + K^T z\right] \tag{4.20}$$

(which is so-called *linearizing control law* and is an IFC) where $K \in \mathbb{R}^\rho$ is chosen in such way that the polynomial $P_\rho(s) = s^\rho + k_\rho s^{\rho-1} + \ldots + k_2 s + k_1 = 0$ is Hurwitz (this is, $P_\rho(s)$ has all its roots in the open left-hand complex plane). For the construction of the canonical form (4.19), and the linearizing control law (4.20), necessary and sufficient conditions are given in [16], [17].

Under the linearizing control law, the closed-loop system (formed by systems (4.19) and (4.20)) can be written in the following form

$$\begin{aligned} \dot{z} &= Az \\ \dot{v} &= \varsigma(z,v;t) \end{aligned} \tag{4.21}$$

where $A \in \mathbb{R}^{\rho \times \rho}$ is the ρ-dimensional companion matrix whose characteristic equation is given by the a Hurwitz polynomial $P_\rho(s)$. In this manner, the following results can be stated.

Lemma 4.6. (Isidori, [17]) Let us assume that there exists a smallest integer ρ such that: (i) $L_g L_f^i h(x) = 0$, i = 1,2,..., ρ-2 and (ii) $L_g L_f^{\rho-1} h(x) \neq 0$, where $L_f^\rho h(x) := L_F^\rho(L_F^{\rho-1} h(x))$, $L_F^0 h(x) := h(x)$ and, by definition, $L_F h(x) := (\partial h(x)/\partial x)F(x)$ is the Lie derivative of $h(x)$ along $F(x)$. This implies that there exists a coordinate transformation $(z,v) = \Phi(x)$ such that the system (4.17) can be globally transformed into the canonical form

$$\begin{aligned} \dot{z}_i &= z_{i+1}; \quad i = 1,2,\ldots,\rho - 1 \\ \dot{z}_\rho &= \alpha(z,v) + \beta(z,v)u \\ \dot{v} &= \zeta(z,v) \\ y &= z_1 \end{aligned} \tag{4.22}$$

where $\alpha(z,v) = L_F^\rho h(x)$, $\beta(z,v) = L_g L_F^{\rho-1} h(x)$, and $v \in \mathbb{R}^{n-\rho}$ denotes the unobservable states, $\zeta(z,v) = [\varphi_{n-\rho}, \varphi_{n-\rho+1}, \ldots, \varphi_n]$ is such that the coordinate transformation $\Phi(x) = [h(x), L_F h(x), \ldots, L_F^{\rho-1} h(x), \varphi_{n-\rho}, \varphi_{n-\rho+1}, \ldots, \varphi_n]$ is globally invertible. Therefore if system (4.22) is minimum phase, i.e., the zero dynamics $\dot{v} = \zeta(0,v)$ is stable, there exists a state feedback controller $u(x) := -(L_F^\rho h(x) + v(x))/L_g L_F^{\rho-1} h(x)$, where $v = K^T \Phi^{-1}(z)$ and $K > 0$ are such that it is asymptotically stable at the origin.

Remark 4.7. Geometrical control theory is well developed and classical books have been published [16], [17]. Thus, relative degree is obtained from the following algorithm. Let $y = h(x)$ be the system output where $h(x)$ is a smooth function. Differentiating the output respect to time we obtain $\dot{y} = (\partial h(x)/\partial x)\dot{x}$ where $\dot{x} = f(x) + g(x)u$ which implies that, by definition of Lie derivative, $\dot{y} = L_F h(x) + L_g h(x) u$. If $L_g h(x) = 0$ (which can be interpreted as the control input does not affects the system output y) for any

$x \in \mathbb{R}^n$, one must differentiate the system output again to get $\ddot{y} = L_F^2 h(x) + L_g L_F h(x) u$. Once again if $L_g L_F h(x) = 0$ for any $x \in \mathbb{R}^n$ the system output should be differentiated respect to time until conditions (i) and (ii) be satisfied. After that one can defined the coordinate transformation $z_{i+1} := L_F^i h(x) = y^{(i)}$ to get system (4.22).

Remark 4.8. It must be pointed out the following: (a) if ρ = n the transformed system (eq. (4.19)) is so-called *fully-linearizable nonlinear system* (FLNS), and (b) if ρ < n the system (4.19) is called *partially-linearizable nonlinear system* (PLNS). In addition, if the dynamical subsystem $\dot{\mathbf{v}} = \varsigma(0, \mathbf{v}) = \varsigma(0,v)$ is asymptotically stable, we will say that the system (4.19) is minimum-phase [27]. Finally, the linearizing control law (4.20) cannot be directly implemented due to it requires information about the nonlinear functions $\alpha(z,v;\pi_3)$ and $\beta(x,v;\pi_4)$, which are uncertain. Hence, the feedback (4.19) must be modified. In what follows, we present the proposed feedback scheme.

Exercise 4.9 *Synchronization of two Rössler systems.* Consider two Rössler systems given by

$$\begin{aligned} \dot{x}_1 &= -(x_2 + x_3) \\ \dot{x}_2 &= x_1 + a x_2 \\ \dot{x}_3 &= 0.2 + x_3(x_1 - 10) \end{aligned}$$

with different initial conditions and $a_S = 1.03 a_M$ (different parameters). Obtain the controller (4.20) such that: (a) synchronize both systems considering an input vector g = $[0\ 0\ 1]^T$ and an output function $y = x_1$. (b) synchronize both systems considering the same input vector but the output function $y = x_2$. Finally, (c) Compare the resulting controllers and state the differences between the synchronous behavior and both controllers.

4.2.2 Synchronization against Uncertain Vector Fields

Since the synchronization problem can be seem as a control one, in what follows, we state a robust adaptive stabilization technique with least prior knowledge which will be used to stabilize the system (4.19) at the origin. The proposed feedback has an adaptive structure and a dynamic compensator where a few control parameters are required to achieve the synchronization (low-parameterized adaptive feedback scheme). The synchronization scheme is based on the control theory discussed in the above section. It must be pointed out that if one is able to stabilize system (4.19) without knowledge about $\beta(z,v;\pi_4)$ and $\alpha(z,v;\pi_3)$, one can stabilize the uncertain system (4.17). Hence, synchronization can be attained.

Now, let us assume the following:

S.4.2.1) The dynamics of the error synchronization (see eq. 4.17) can be transformed into the canonical form (4.19), *i.e.*, there exists a diffeomorphic transformation of coordinates given by eq. (4.18).

S.4.2.2) Only, the system output of the transformed system, $y = z_1$, is measurable.

S.4.2.3) In the system (4.16), the vector field $\Delta F(x,t;p)$ and the nonlinear function $\Delta T(t;p_2)$ are unknown. Consequently, the nonlinear functions $\alpha(z,v;\pi_3)$ and $\beta(z,v;\pi_4)$ are uncertain. However, an estimate $\beta_E(z)$ of $\beta(z,v;\pi_4)$, satisfying $\text{sign}(\beta(z,v;\pi_4)) = \text{sign}(\beta_E(z))$, is available for feedback.

S.4.4) The system (4.16) is *minimum-phase* (see [27]).

Comments regarding the above assumptions are in order:

(a) For Assumption *S.4.2.1* Several systems subjected to chaotic synchronization can be transformed into the canonical form (4.19). For example, the Chua circuit dynamical system can be transformed into the canonical form with a relative degree, $\rho < n$. On the other hand, the nonautonomous second-order chaotic systems like the Duffing oscillator is a FLNS. Such oscillators can be written as $\dot{x}_1 = x_2, \dot{x}_2 = f(x) + u(x)$ where $u(x)$ is the control command.

(b) The Assumption *S.4.2.2* is realistic because in most cases only one state is available for feedback from the coding (master or drive system) as well as the decoding (slave or response system) circuit. For instance, in the secure communication case only the transmitted signal, $x_{1,M}$, and receiver signal, $x_{1,S}$, are available for feedback from measurements. Other example can be found in neuron synchronization where the master neuron transmits a scalar signal. The slave neuron tracks the signal of the master neuron.

(c) Concerning Assumption *S.4.2.3*, we claim that it is a general and practical situation because the terms $\Delta F(x,t;\pi_1)$ and $\Delta T(t;\pi_2)$ involve the differences in the master as well as slave systems. The source of such uncertainties are: (i) unknown values of the model parameters or time-varying parameters, (ii) since the master and slave systems are chaotic; their trajectories depend on the initial conditions which are often unknown and, (iii) structural differences between models of the master and slave system.

(d) Regarding the Assumption *S.4.2.4*, *minimum-phase* implies that the zero dynamics $\dot{\mathbf{v}} = \zeta(0,\mathbf{v},t)$ converges to an attractor. In other words, the closed-loop system is internally stable. Form the control theory viewpoint this is the stronger assumption. However, several interesting chaotic systems satisfy the internal stability assumption.

Since $\Delta F(x;\pi_1)$ and $\Delta T(t;\pi_2)$ are uncertain (*Assumption A.3*), the Eq. (4.18) is an *uncertain* nonlinear change of coordinates, hence $\alpha(z,v;\pi_3)$ and $\beta(z,v;\pi_4)$ in the transformed system (4.19) are also unknown. The idea to deal with the uncertain terms $\alpha(z,v;\pi_3)$ and $\beta(z,v;\pi_4)$ is to lump them into a new function which can be interpreted as a *new observable state*. By an *observable state* we mean that the dynamics of such state can be reconstructed from on-line measurements (for example, $y = z_1$).

According to *S.4.2.4*), the functions $\alpha(z,v;\pi_3)$ and $\beta(z,v;\pi_4)$ are uncertain. Thus, let us define $\delta(z,v;\pi_4) = \beta(z,v;\pi_4) - \beta_E(z)$, and $\Theta(z,v,u;\pi) = \alpha(z,v;\pi_3) + \delta(z,v;\pi_4)u$. Simple algebraic manipulations yield the following expression for the transformed system

$$\begin{aligned} \dot{z}_i &= z_{i+1}; \qquad 1 < i \leq \rho \\ \dot{z}_\rho &= \Theta(z, v, u; \pi) + \beta_E(z)u \\ \dot{v} &= \varsigma(z, v) \end{aligned} \tag{4.23}$$

where $\Theta(z,v,u;\pi)$ is a term which lumps the uncertainties. Now, let us define $\eta = \eta(t) \equiv \Theta(z(t),v(t),u(t);\pi)$. In this way the system (4.23) can be rewritten in the following extended form

$$\begin{aligned} \dot{z}_i &= z_{i+1}; \qquad 1 < i \leq \rho \\ \dot{z}_\rho &= \eta + \beta_E(z)u \\ \dot{\eta} &= \Xi(z, \eta, u; \pi) \\ \dot{v} &= \varsigma(z, v) \end{aligned} \tag{4.24}$$

where $\Xi(z,v,\eta,u;\pi) = \sum_{k=1}^{\rho-1} z_{k+1}\partial\Theta(z,v,u;\pi) + [\eta + \beta_E(z)u]\partial_\rho\Theta(z,v,u;\pi) + \delta(z,v;\pi)\dot{u} + \partial_t\delta(z,v;\pi)u + \partial_v\Theta(z,v,u;\pi)\zeta(z,v,\eta)$, and $\partial_k\Theta(z,v,u;\pi) = \partial\Theta(z,v,u;\pi)/\partial x_k$, $k = 1,2,\ldots,\rho$.

It is straightforward to prove that $\Psi(z,\eta,v,u;\pi) = \eta - \Theta(z,v,u;\pi)$ is a first-integral of the system (4.24) [29]. This means that along the trajectories of system (4.24), this is $d\Psi(z,\eta,v,u;\pi)/dt = 0$ for all $t \geq 0$ and any differentiable control $u(t)$. Consequently, the states representation (4.24) is dynamically equivalent to the system (4.23) (and consequently to the system (4.19)), as long as initial conditions satisfy $\Psi(z_0,\eta_0,v_0,u_0;\pi) = 0$. This is, the *augmented state* η provides the dynamics of the uncertain function $\Theta(z,v,u;\pi)$ which involves the modelling differences, uncertain parameters and the unknown external disturbances.

On the other hand, there are two important remarks concerning the system (4.24). (a) A geometrical interpretation of the states representation (4.24) is as follows. If $z(t)$ is a solution of the system (4.23) with initial conditions (z_0,u_0) and $(z(t),\eta(t))$ is a solution to eq. (4.24) with initials conditions (z_0,η_0,u_0), then the equality $\Psi(z_0,\eta_0,v_0,u_0;\pi) = 0$ and the condition $d\Psi(z,\eta,v,u;\pi)/dt = 0$ imply that the solution of (4.23) is a projection of the solution of (4.24). (b) A feature of eq. (4.24) is that the *uncertainties have been lumped in an uncertain function* $\Theta(z,v,u;\pi)$ *which can be estimated by means of a nonmeasurable but observable state* η. Furthermore, if one is able to stabilize the system (4.24) without making use of the constraint $\Psi(z_0,\eta_0,v_0,u_0;\pi) = 0$, one would be able to stabilize the system (4.19) and its equivalent system (4.16).

Now, let us consider the following linearizing control law to stabilize the system (4.24)

$$u = \frac{1}{\beta_E(z)}(-\eta + K^T z) \tag{4.25}$$

where $K \in \mathbb{R}^\rho$ is chosen as in eq. (4.20). If the initial conditions satisfy $\Psi(z_0,v_0,\eta_0,u_0;\pi) = 0$, then under (4.25) the states (z,η) converges exponentially to zero. Convergence of z to zero follows from the fact that the closed-loop system is minimum-phase. Since $\Delta F(x;\pi_1)$ is smooth, $\alpha(z,v;\pi_3)$ is also smooth. Then, the control dynamics is given by $\dot{u} = (-\Xi(z,v,\eta,u;\pi) + K^T\dot{z})/\beta_E$. Since $\Xi(z,v,\eta,u;\pi)$ is a smooth function, $\dot{u}$ is also a smooth function. Consequently, $\Psi(z,\eta,v,u;\pi) = \eta - \Theta(z,v,u;\pi)$ is a *first-integral* of the closed-loop system [15]. Then, from eq. (4.24), the augmented state becomes: $\eta = K^T z - \beta_E(z)u$ hence the augmented state η is bounded and its dynamics is also bounded.

Nevertheless, the linearizing feedback control law (4.25) is not physically realizable because it requires measurements of the states $z(t)$ and the uncertain state $\eta(t)$. As it has been established in [25], the problem of estimating (z,η) can be addressed by using a high-gain observer. Thus, *the dynamics of the states z and η can be reconstructed from measurements of the output $y = z_1$ in the following way*

$$\begin{aligned}
\dot{\hat{z}}_i &= z_{i+1} + L^i\kappa_i(z_1 - \hat{z}_1) \quad 1 \leq i \leq \rho - 1\\
\dot{\hat{z}}_\rho &= \hat{\eta} + \beta_E(z)u + L^\rho\kappa_\rho(z_1 - \hat{z}_1)\\
\dot{\hat{\eta}} &= L^{\rho+1}\kappa_{\rho+1}(z_1 - \hat{z}_1)
\end{aligned} \tag{4.26}$$

where the κ'_js, $j = 1,2,\ldots,\rho+1$, are chosen such that the polynomial $P_\kappa(s) = s^{\rho+1} + \kappa_{\rho+1}s^{\rho-1} + \ldots + \kappa_2 s + \kappa_1 = 0$ is Hurwitz. The high-gain parameter, L, can be interpreted as the uncertainties estimation rate and it is the unique tuning parameter.

Thus, the linearizing feedback control law with uncertainty estimation becomes

$$u = \frac{1}{\beta_E(\hat{z})}(-\hat{\eta} + K^T\hat{z}) \tag{4.27}$$

Notice that the linearizing control law (4.27) only uses estimates of $\Theta(z,v,u;\pi)$ (by means of $\hat{\eta}$), $\hat{z}$ and $\beta_E(z)$ which are provided by estimator (4.26). Thus, the modified feedback control law is given by the dynamic compensator (4.26) and the linearizing control law (4.27). The controller (4.26),(4.27) yields practical stabilization, *i.e.*, the system (4.24) converges to a ball B' whose radius is proportional to L^{-1}. That is, $x \to B'$ as $t \to \infty$ with $r(B') = O(L^{-1})$.

From a practical viewpoint, the above statement implies that the synchronization error can be made as small as desired by taking large values of the parameter $L > 0$. This result has a lot of analogy with those in Mettin et. al. work [28]. In fact, exact synchronization (*i.e.*, zero synchronization error) is not possible due to the difference in the actual and estimated dynamics. These comments will be illustrated via numerical simulation below.

Note that since $\Theta(z,v,u;\pi)$ and $\delta(z,v;\pi)u$ are uncertain, the function $\Xi(z,\eta,v,u;\pi)$ is correspondingly unknown. Thus, such term was not used in the construction of the observer (4.26). This feature yields a low order parameterization (only a tuning parameter is required) to the dynamic compensator of the adaptive strategy. This is an advantage respect to previously reported control schemes.

Exercise 4.10. *Synchronization of two Rössler systems.* Repeat Exercise 4.9, but construct system (4.26) and the controller (4.27) and compare the performance between the here obtained controllers and those obtained in Exercise 4.9. Also discuss the execution of the controllers.

4.2.3 Complete Practical Synchronization of Chaotic Systems

Essentially, in this section, we show that the use of the proposed algorithm yields to continuous feedback such that the synchronization can be achieved even if the master and slave model are non-identical. To this end, the Chua oscillator has been considered as illustrative example. In fact, the Chua oscillator exhibits a family of chaotic attractors and it can be easily implemented in hardware [30]. Several works have been reported on the synchronization of the Chua circuit via feedback and nonfeedback control. Thus, we use the Chua oscillator to illustrate how our feedback strategy could be used in experimental implementations. Morever, experimental implementation is not hard to realize (see [31] for possible experimental setup). The Chua oscillators is an electronic circuit which consists of one linear inductor, two linear resistors, two linear capacitors and a nonlinear resistor which is so-called Chua diode [3]. The synchronization problem of this circuit has been widely studied. Experimental implementation of adaptive proposed strategies have been reported (see [5], [22] and references therein). Although in this section the implementation is performed via numerical simulations, it is not hard to see that the physical application of the proposed adaptive feedback can be performed. The dimensionless model of the model becomes [3]

$$\begin{aligned} \dot{x}_{1,j} &= \alpha\left(x_{2,j} - x_{1,j} - f_c(x_{1,j})\right) \\ x_{2,j} &= x_{1,j} - x_{2,j} + x_{3,j} \\ \dot{x}_{3,j} &= -\beta x_{2,j} \end{aligned} \tag{4.28}$$

where the subscript, j = M,S, indicates the master and slave system, $f_C(x_{1,j}) = a_1 x_{1,j} + \frac{1}{2}(a_1 - a_2)(|x_{1,j} + 1| - |x_{1,j} - 1|)$, $x = (x_{1,j}, x_{2,j}, x_{3,j})$ is a state vector. $x_{1,j}$ represents, in some sense, the voltage cross the first linear capacitor, $x_{2,j}$ is associated with voltage cross the second capacitor and $x_{3,j}$ is the current through the linear inductor [3], respectively.

Assume that (i) the control command is a signal injected in the voltage across the first linear capacitor of the slave circuit (*i.e.*, according to Assumption *A.2*, $y = x_1$ is the measured state). If the measured state is the voltage cross the first linear capacitor, the relative degree of the Chua system is $\rho = 1$ and the zero dynamics is stable which implies that minimum-phase assumption is satisfied (see next section). Now, let us define $x_i = x_{i,S} - x_{i,M}$. Then, the dynamics of the synchronization error can be written as follows

$$\begin{aligned} \dot{x}_1 &= \Delta f_1(x) - u \\ \dot{x}_2 &= \Delta f_2(x) \\ \dot{x}_3 &= \Delta f_2(x) \\ y &= x_1 \end{aligned} \tag{4.29}$$

where $\Delta f(x) = [\Delta f_1(x), \Delta f_2(x), \Delta f_3(x)]^T$ is an uncertain, but smooth, vector field. Then, by taking the coordinates exchange, $z = y = x_1$, the system (4.29) can be transformed into the canonical form (4.19) where $\alpha(z,v;\pi_3) = \alpha_M x_2 - \alpha_M x_1 - u + \Delta\alpha x_{2,s} - \Delta\alpha x_{1,s} - \Delta\alpha\Delta f_C(\Delta x_1)$ is an uncertain nonlinear function ($\alpha(z,v;\pi_3)$ contains the differences, Δ, between master and slave oscillators. Moreover, the value of the master parameter, α_M, is unknown) and $\beta(z,v;\pi_4)$ is, in particular, an unknown constant. Thus, defining $\beta_E(z) = 1.0$, $\delta(z,v;\pi) = \beta(z,v;\pi_4) - \beta_E(z)$, $\Theta(z,v,u;\pi) = \alpha(z,v;\pi_3) + \delta(z,v;\pi_4)u$ and $\eta = \Theta(z,v,u;\pi)$, the extended system (4.24) takes the form

$$\begin{aligned} \dot{z}_\rho &= \eta - u \\ \dot{\eta} &= \Xi(z,\eta,u;\pi) \\ \dot{v} &= \varsigma(z,v) \end{aligned} \tag{4.30}$$

where $v = (x_2, x_3)$ is the zero dynamics and $\Xi(z,v,\eta,u;\pi)$ is a nonlinear function which is given as above. Hence, the controller (4.26),(4.27) becomes

$$\begin{aligned} \dot{\hat{z}} &= \hat{z} + L\kappa_1(z - \hat{z}) \\ \dot{\hat{\eta}} &= L^2\kappa_2(z - \hat{z}) \\ u &= -\hat{\eta} + k\hat{z} \end{aligned} \tag{4.31}$$

Figure 4.5 shows the stabilization of the synchronization error dynamics and both Chua oscillators behaves in synchronous way. The control gain value k was chosen as 7.0. The constants κ'_is were chosen such that the roots of $P_\kappa(s) = 0$ are located at -1.0. The high-gain estimation parameter value is L = 10. The controller is activated at t = 55.0 (for t < 55 u = 0.0). In Figure 4.5, the master parameters values are the following: $\gamma_{1,M}$ = 10.0, $\gamma_{2,M}$ = 14.87, $a_{1,M}$ = -1.27 and $b_{1,M}$ = - 0.68 and the slave parameters values were chosen as $\gamma_{1,S}$ = 9.0, $\gamma_{2,S}$ = 15.0, $a_{2,S}$ = -1.5 and $b_{1,S}$ = - 0.8. Notice that the synchronization error is stabilized at the origin by the controller in spite of the fact that both master and slave circuits have different parameters values. An important feature is the following: *Despite the control command u is only acting on the state z,* $v \in \mathbb{R}^2$ *is also stabilized.*

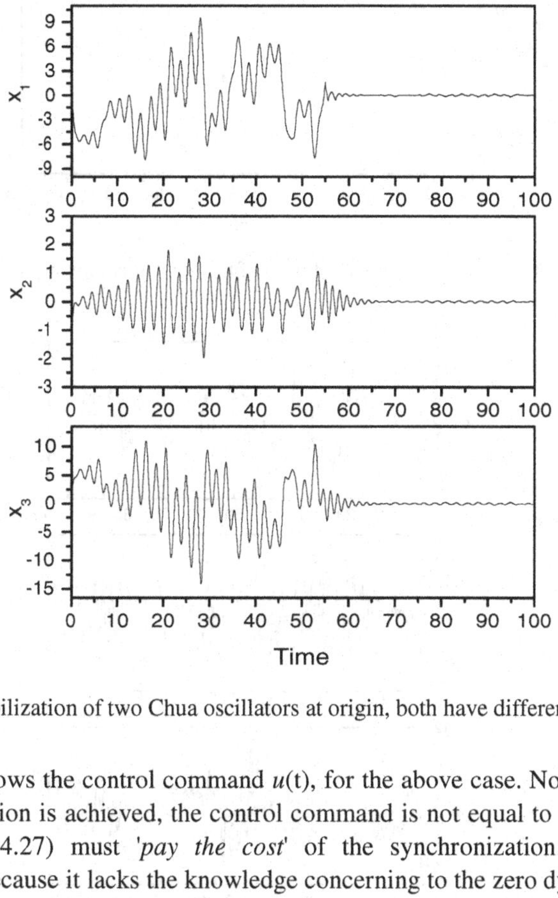

Fig. 4.5. Stabilization of two Chua oscillators at origin, both have different parameters

Figure 4.6 shows the control command $u(t)$, for the above case. Note that although the synchronization is achieved, the control command is not equal to zero due to that the controller (4.27) must *'pay the cost'* of the synchronization with dynamic compensation because it lacks the knowledge concerning to the zero dynamics. Figure 4.7 shows the master and slave states for the case presented, dashed line represents the slave system and continuous line the master. Since the synchronization error is stabilized at the origin, the master and slave trajectories are synchronous.

Potential applications of the proposed robust synchronization technique are: (a) secure signal transmission and, (b) synchronization of inhomogeneous chain of oscillators. Due to secure signal transmission is a complex topic, it will be studied in the next section. Finally, the synchronization of an inhomogeneous chain of chaotic oscillator is presented in this part.

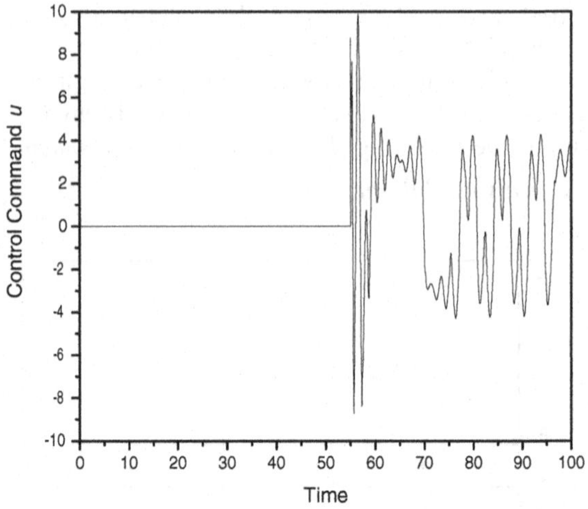

Fig. 4.6. Dynamic evolution of the control command u

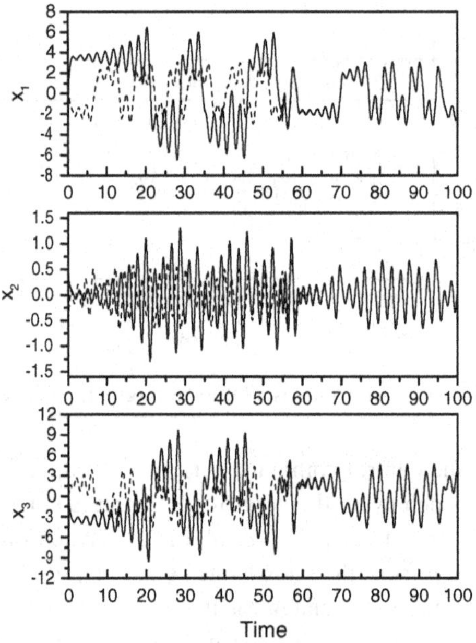

Fig. 4.7. Synchronization of two Chua oscillators with different parameters, all states behave in synchronous way

4.2.4 Synchronization of an Inhomogeneous Chain of Oscillators

The synchronization of an inhomogeneous chain of oscillators can be applied to the multipodal animal gait or cellular nonlinear networks [32]. It has been reported that such systems can be modelled by second-order driven oscillators (for instance van der Pol [9]). In addition, it is well known that such oscillators, under certain conditions, can exhibit chaos. Hence the robust synchronization is required.

Now, consider a chain of unidirectionally coupled chaotic systems where the oscillations in the slave subsystems must be synchronized with those of the master subsystem. Let us assume that the chaotic systems in the chain of oscillators are given by N non-identical Chua's circuits (one master and N-1 slave oscillators). In this way, a PLNS can be used in order to reduce the number of differential equation to be solved in the synchronization scheme. Such chain structure has the form of *clusters in lattices of chaotic neurons* [33].

Figure 4.8 shows the synchronization in a chain with $N = 4$ oscillators (1 master and 3 slaves oscillators). The parameters were chosen as follows: (a) *Master*: $\gamma_{1,M} = 10.0$, $\gamma_{2,M} = 14.87$, $a_{1,M} = -1.27$ and $b_{1,M} = -0.68$; (b) *Slave 1*: $\gamma_{1,S1} = 9.0$, $\gamma_{2,S1} = 15.0$, $a_{1,S1} = -1.5$ and $b_{1,S1} = -0.8$; (c) *Slave 2*: $\gamma_{1,S2} = 9.5$, $\gamma_{2,S2} = 14.5$, $a_{1,S2} = -1.3$ and $b_{1,S2} = -0.65$; (d) *Slave 3*: $\gamma_{1,S3} = 9.0$, $\gamma_{2,S3} = 14.5$, $a_{1,S3} = -1.3$ and $b_{1,S3} = -0.7$. The control and observer parameters values were chosen as in Section 3. The control is activated at $t = 20$ (*i.e.*, for $t < 20$ $u = 0$). Note that the proposed control technique allows the practical synchronization of all oscillators of the chain despite uncertainties in the parameters of the oscillators.

Figure 4.9 shows the relation between the measured states of each oscillator of the inhomogeneous chain. Note that the phase of the slave systems is locked, which is a common measure of the degree of synchronization. Since the stabilization of the synchronization error around an arbitrarily small neighbourhood of the origin is guaranteed (see next section), the locking of the phase was expected.

The propose approach is based on geometrical control theory to provide a systematic procedure to design an adaptive synchronization strategies. The design

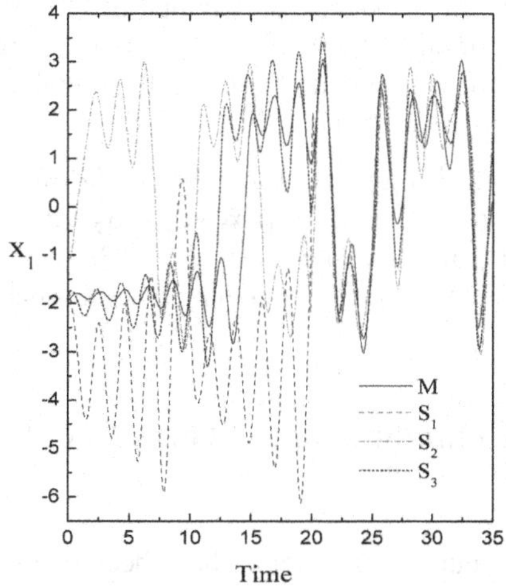

Fig. 4.8. Synchronization of three different Chua oscillators, continuous line represents the master system

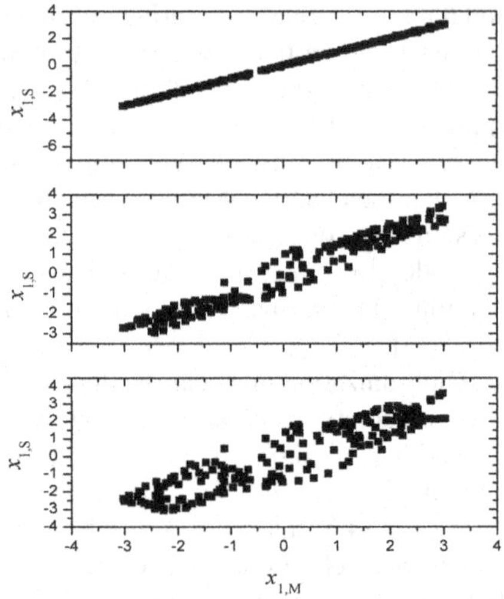

Fig. 4.9. Phase locking for the three slave oscillators

algorithm yields continuous feedback which allows handling realistic situations such as unmeasured states and uncertainties. So, synchronization of unequal systems can be attained. The resulting synchronization scheme is obtained by posing a control problem in the synchronization error dynamics. The control problem consists in the stabilization at the origin of the synchronization error system. The proposed scheme comprises a linearizing-like control law and a state/uncertainty estimator. A central feature of our approach is that uncertainties of the underlying vector field are lumped in an extended state whose dynamics is reconstructed from measurements of the system output. In this way, the robust asymptotic control scheme allows the stabilization of the synchronization error system at the origin. Consequently, the synchronization of both master and slave oscillators is attained. In fact, the phase of the master and slave system is locked (see Figure 4.9).

Exercise 4.11. *Synchronization of two hyperchaotic Rössler systems.* Consider two hyperchaotic systems given by $\dot{x}_1 = -(x_2 + x_3)$, $\dot{x}_2 = x_1 + ax_2 + x_4$, $\dot{x}_3 = 3 + x_3x_1$, $\dot{x}_4 = -0.5x_3 + 0.005x_4$, assume different initial conditions and $a_S = 1.03a_M$. Obtain the controller (4.26) and (4.27), that synchronize both systems assuming that the input vector g = [0 0 1 0] and the output function $y = x_2$.

4.3 Chaos-Based Communication Scheme Via Robust Asymptotic Feedback

Several chaos-based communication schemes have been published. Among these, two basic configurations can be identified: (i) An approach consists of the addition of the message signal to the chaotic carrier (transmitted signal) which is sent to the receiver. That is, the master system (the so-called drive system or transmitter) comprises the full-state model whereas the slave system (the so-called response system or receiver) is composed by a reduced model (homogeneous synchronization, see for instance references [22] and [33]). (ii) Another transmitter/receiver design is based on the full-state model of the driving and response systems [34]. That is, both drive and response systems are represented by dynamical systems of the same order. The homogeneous synchronization configuration has been recently addressed via parameter modulation. Kocarev and Parlitz have proposed a generalization of these approaches which extends the capabilities for constructing synchronized systems [35]. Their approach enables the message signal to be integrated as a driving signal. However, the message signal can be recovered only under ideal conditions [35].

The synchronization of non-identical chaotic systems and its usage in transmitter/receiver configurations is an interesting problem. There are two main objectives to be considered here. The first one is to test the security level of communications based on synchronization of chaotic systems. The point is, can a response system unmask a message signal in spite of the drive system is unknown?. The synchronization of two strictly different chaotic systems is possible [25]. Hence, a response system whose dynamical model is not similar than the drive system can reconstruct the message signal. The latter objective is related with the dynamic reconstruction of the master dynamics (drive system). This problem can be addressed from the control theory point of view, where two basic ideas can be identified:

(a) *Construct an observer.* A dynamical system (which is called an *observer*) is constructed from the transmitter model in such a way that, it yields the dynamical reconstruction of the transmitter states. In this way, the observer results in a response system which can be synchronized in spite of unknown initial conditions of the drive system [35], [36]. A standard approach to solve the observer problem is to use a receiver that is a copy of the transmitter (with unknown initial conditions). The copy of the transmitter is modified with a term depending on the difference between the drive and response system [37]. Such an additional term serves to attenuate the dynamics of the synchronization error states. Therefore this procedure requires *a priori* knowledge of the drive system model. Moreover, although the procedure could be shown to be successful in many instances, no global synchronization can be claimed. However, the main idea behind the observer construction is fundamentally deterministic. That is, according to the Takens theorem [38], there are certain conditions for which the dynamics of a system can be embedded in a finite-dimensional space. In this sense, the deterministic systems have an *observability property*. The observability property means that the history of the transmitted signal contains all the information required to reconstruct the states of the drive system. In such a case it is said that the drive system is *observable* [38].

(b) *Design a feedback scheme.* The main idea here is to design a control loop in the receiver system. The feedback steers the states of the synchronization error system to the origin; *i.e*, the states of the response system *track* the states of the drive system. The feedback results in a coupling force between transmitter and receiver in such way that discrepancy between the driving and response systems is controlled. In this sense, the synchronization problem can be interpreted as a chaos suppression problem. Since 90's, chaos suppression has been intensively studied, some relevant references can be mentioned [39], [40], [41], [42]. It has been shown that, under feedback control, the chaos control can be attained [15], [41]. This means that the dynamics of the synchronization error can be led to origin in spite of the driving and response models are non-identical (and even if they are strictly different [25]). Since the chaos suppression can be achieved under uncertain vector fields and unknown initial conditions, the feedback yields the chaos synchronization. As a consequence, this interpretation can also be applied to secure communication. Nevertheless, a minimum phase condition is often assumed. Although, this condition can be satisfied in many situations, minimum phase is a strong assumption.

In this section, the chaos synchronization problem and its application to communication is addressed from a control theory perspective. In particular, the synchronization problem is interpreted as a stabilization one. The goal is to stabilize, at the origin, the discrepancy between the driving and response systems. Discrepancy is defined as the dynamical differences between driving and response systems and includes: (i) model mismatches, which means that the model of the drive system could not be the same that response system. (ii) unknown initial conditions, which implies that the time series of the transmitter cannot be equal than for the receiver. (iii) parametric uncertainty, which means that the receiver circuit could be constructed with inaccuracies.

In order to design the stabilizing feedback the discrepancy system is transformed into a canonical form via geometrical tools [17]. The main idea behind our proposal is, departing from the discrepancy system, to construct an extended nonlinear system which should be dynamically equivalent to the canonical representation. In this way, the discrepancy is lumped in a nonlinear function, which is rewritten into the extended nonlinear system as a state variable. By, using the results reported by Teel and Praly [43], an observer can be constructed to get an estimated value of the lumping nonlinear function via the augmented state variable. In fact, this procedure has been used to perform synchronization of strictly different chaotic oscillators [25]. In this section, the procedure presented in Section 4.2.2 is used to develop a secure communication scheme.

4.3.1 Problem Statement

Consider the following nonlinear system

$$\begin{aligned} \dot{x}_S &= f_S(x_S;\pi_S) + g_S(x_S)u \\ y_S &= h(x_S) \end{aligned} \tag{4.32}$$

where $x_M \in \mathbb{R}^n$, $\pi_M \in \mathbb{R}^P$ is a parameter set, $s_m \in \mathbb{R}$ is a bounded known function and denotes a message signal (which is injected into the nonlinear system), $f_M:\mathbb{R}^n \to \mathbb{R}^n$ and $g_M:\mathbb{R}^n \to \mathbb{R}^n$ are smooth vector fields, $y_M \in \mathbb{R}$ and $h(x_M)$ is a smooth function, that determines the *transmitted* states.

Remark 4.12. The message signal s_m can be interpreted as an input signal, i.e., there exist a signal s_m with suitable frequency and amplitude such that the dynamics of system (4.32) is chaotic. Note that message signal can be represented by: (i) a modulated parameter of the drive system; in such a case it is said that the message signal is *demodulated* by response system or (ii) a smooth time function which is not a parameter of the drive system; in such a case the message signal is *decoded* by response system.

Now, let us consider the following nonlinear system

$$\begin{aligned} \dot{x}_S &= f_S(x_S;\pi_S) + g_S(x_S)u \\ y_S &= h(x_S) \end{aligned} \tag{4.33}$$

where $x_S \in \mathbb{R}^n$, $\pi_S \in \mathbb{R}^q$ is a parameter set, $f_S:\mathbb{R}^n \to \mathbb{R}^n$ and $g_S:\mathbb{R}^n \to \mathbb{R}^n$ are smooth vector fields, the input $u \in \mathbb{R}$ denotes the demodulated signal. $y_S \in \mathbb{R}$ and $h(x_S)$ is a smooth function which determines the *comparison* states. The comparison states are chosen in such a way that they are the same than the transmitted states, *i.e.*, output of the response system y_S is chosen equal to the transmitted state y_M which does not means that channel is perfect. Indeed, it means that same state at drive and response system are available for feedback. Henceforth security level of communication scheme is fixed by comparison states.

Remark 4.13. The nonlinear systems (4.32) and (4.33) have the same order. From the synchronization (chaos theory) point of view, this assumption is strong. However, it is a standard consideration for the signal transmission case. In addition note that vector field $g_S(x_S)$ represents the way as the input u is entered into system (4.33).

By defining $x_i := x_{i,M} - x_{i,S}$, the following uncertain dynamical system can be obtained

$$\begin{aligned} \dot{x} &= \Delta f(x;\pi) + g_M(x_M)s_m - g_S(x_S)u \\ y &= h(x) \end{aligned} \tag{4.34}$$

where $x \in \mathbb{R}^n$, $\pi \in \mathbb{R}^r$ is a parameter vector and Δf: $\mathbb{R}^n \to \mathbb{R}^n$ is a vector field defined by $\Delta f(x,\pi) := f_M(x_M;\pi_M) - f_S(x_S;\pi_S)$, $h(x)$ is a smooth function which determines the difference between the *transmitted* and *comparison* states. Note that models of the drive and response systems are assumed non-identical, hence $\Delta f(x,\pi)$ is an uncertain vector field. Finally, the term $g(x_M)s_m$ is unknown.

Definition 4.14. (*Practical synchronizability*). It is said that the chaotic systems (4.32) and (4.33) are practically synchronized if, for a compact set of initial conditions, $X_0 \subseteq \mathbb{R}^n$, and for each pair of compact sets (U_P,U_{PS}), which are neighborhoods of the origin with $U_P \subset U_{PS}$, there exists a locally Lipschitz function, $u = a_1\|x\|$, where a_1 is a positive constant, such that the solution of $x(t) = (x_M(t) - x_S(t)) \in U_{PS} \subset \mathbb{R}^n$, system (4.34),with initial conditions $x(0) \in X_0$ belongs to U_p for all $t \geq 0$.

Remark 4.15. Since trajectories $x_M(t)$, $x_S(t)$ of uncontrolled systems (4.32) and (4.33) are chaotic hence differences $x_i(t) := x_{i,M}(t) - x_{i,S}(t)$ belongs to any chaotic attractor [25]; that is trajectories $x(t)$ start at initial condition $x_i(0) := (x_{i,M}(0) - x_{i,S}(0)) \in X_0$ and converge to the set U_P which is not necessarily equal to X_0. Moreover, if systems (4.32) and (4.33) can be synchronized then trajectories $x(t)$ will converge to any neighborhood U_{PS} of the origin which is contained in the attractor set U_P. Finally, Definition 4.14 includes several kinds of synchronization: i) Complete exact synchronization (CES) (where $\|x_S(t) - x_M(t)\| \equiv 0$ for all $t \geq 0$), ii) Complete inexact synchronization (where $\|x_S(t) - x_M(t)\| \approx 0$ for all $t \geq 0$), iii) Partial synchronization (where at least for one state $x_i(t)$, for any $i \leq n$, $\|x_M(t) - x_S(t)\| \neq 0$) and iv) Almost synchronization (where only the phase of the drive system is similar to the response system but the amplitude is different).

In order to show the communication, we have made the following assumptions:

Assumption A.1) The order of the drive and response system is the same.

Assumption A.2) There are uncertain model mismatches between the transmitter and the receiver; *i.e.*, $f_M(x_M) \neq f_S(x_S)$ which implies that the discrepancy model is unknown.

Assumption A.3) $g_S(x_S) \neq g_M(x_M)$ where $g_S(x_S)$ is known and bounded away from zero. In addition $\text{sign}(g_S(x_S)) = \text{sign}(g_M(x_M))$.

Assumption A.4) System (4.34) is minimum phase, *i.e.*, the zero dynamics $\dot{v} = \zeta(0,v)$, where $v \in \mathbb{R}^{n-\rho}$, is asymptotically stable.

Some comments about the above assumptions are in order. Assumptions (A.1) and (A.2) are realistic situations. Indeed transmitter and receiver oscillators are often designed with same order and model discrepancy can be expected due to differences between actual values in components of the circuit. In extreme cases, model of transmitter could be strictly different than the receiver. However, recovering of the message signal is desired even in extreme cases. Assumption A.3) implies that if exact synchronization between drive and response system is attained one has that $f_M(x_M) - f_S(x_S) = 0$ and $x_M(t) - x_S(t) \equiv 0$ for all $t \geq 0$ then $g_S(x_S)s_m \equiv g_M(x_M)u$ for all $t \geq 0$, i.e., $u/s_m = g_S(x_S)/g_M(x_M)$ for all $t \geq 0$. According to Assumption A.3) $\text{sign}(g_S(x_S)) = \text{sign}(g_M(x_M))$ the message signal s_m will be recovered with a factor $g_S(x_S)/g_M(x_M)$. Minimum phase supposition is stronger condition which implies that uncontrollable states $v \in \mathbb{R}^{n-\rho}$ of the discrepancy system are asymptotically stable. Fortunately the most chaotic oscillators satisfy this assumption. Now, under the above assumptions, the synchronization problem becomes: *Is there any smooth function $u = u(x)$ such that the uncertain nonlinear system (4.22) is asymptotically stable at the origin and the message signal can be decoded* ?. In this sense the synchronization problem has been interpreted as chaos suppression one, *i.e.*, if $x \to 0 \Rightarrow x_S \to x_M$.

4.3.2 Results on Secure Communication Via Feedback

Definition 4.16. (*Practical semiglobal stabilizability*). Let us consider the affine system $z = f(z) + g(z)u$, where $z \in \mathbb{R}^n$ and $f(z)$ and $g(z)$ are smooth vector fields. The equilibrium $z^* = 0$ of the affine system is said to be semiglobally practically stabilizable by dynamic output feedback if, for each pair of compact sets (U_P, U_{PS}), which are neighbourhoods of the origin with $U_P \subset U_{PS}$, there exists: (i) a locally Lipschitz function $u = u(z)$ and (ii) a pair of compact sets (U_{Po}, U_{PSo}) such that the solution of the affine system $z(t) \in \mathbb{R}^n$ with initial conditions in the basin of attraction in $U_{PS} \times U_{PSo}$ converges to the set $U_P \times U_{Po}$ for all $t \geq 0$.

For sake of clarity in presentation, we should point out that Practical Synchronizability (Definition 4.14) and Practical Semiglobal Stabilizability (Definition 4.16) are very close notions. Indeed, from the viewpoint of the framework theory presented in this paper the practical synchronizability problem is a stabilization problem. The idea is to lead the states of discrepancy system $x(t) \in \mathbb{R}^n$ from initial conditions $x(t = 0)$ in a subset X_0 of $\mathbb{R}^n$ to an arbitrarily small subset $U_P \subset U_{PS}$ containing origin. Several kinds of synchronization can be found when feedback control is used [44], Definition 4.14 and 4.16 comprise all kinds of synchronization.

Theorem 4.17. Let us assume that there exists a coordinate transformation $z = \Phi(x)$ such that the uncertain nonlinear system (4.34) can be transformed into the canonical form (4.22). Now, suppose that there is an estimate $\beta_E(z)$ defined by Lie derivative along $g_S(x_S)$ of the uncertain function $\beta(z,v) := L_g L_F^{\rho-1} h(x)$ such that $\mathrm{sign}(\beta_E(z)) = \mathrm{sign}(\beta(z,v))$ and let us define $\delta(z,v) = \beta(z,v) - \beta_E(z)$, $\Theta(z,v,u) = \alpha(z,v) + \delta(z,v)u$ and $\eta = \Theta(z,v,u)$. Then, there exists an invariant manifold such that the nonlinear system (4.22) can be rewritten in the following form

$$\begin{aligned} \dot{z}_i &= z_{i+1}; \quad i = 1,2,\ldots,\rho-1 \\ \dot{z}_\rho &= \eta + \beta_E(z)u \end{aligned} \tag{4.35}$$

$$\begin{aligned} \dot{\eta} &= \Gamma(z,v,\eta,u,\dot{u}) \\ \dot{v} &= \zeta(z,v) \\ y &= z_1 \end{aligned} \tag{4.36}$$

where $\Gamma(z,v,\eta,u,\dot{u}) = \sum_{i=1}^{\rho-1}[z_{i+1}\partial_i \Theta(z,v,u)] + [\eta + \beta_E(z)]\partial_\rho \Theta(z,v,u) + \delta(z,v)\dot{u} + \partial_v \Theta(z,v,u)$.

Proof. By definition $\delta(z,v) = \beta(z,v) - \beta_E(z)$, $\Theta(z,v,u) = \alpha(z,v) + \delta(z,v)u$ and $\eta = \Theta(z,v,u)$. This implies that there exists a manifold $\Psi(z,\eta,v,u) = \eta - \Theta(z,v,u)$ such that $d\Psi(z,\eta,v,u)/dt = 0$ (i.e., it is a time invariant manifold with boundary) for the initial condition $\Psi_0 = 0$. Since by definition $\Psi(z,\eta,v,u) = \eta - \Theta(z,v,u)$, implies $d\Psi(z,\eta,v,u)/dt = 0$, it means that $\Psi(z,\eta,v,u)$ is the first integral of the nonlinear system (4.35). Therefore the solution of system (4.22) is a projection of the solution of system (4.35). Hence system (4.35) is dynamically equivalent to system (4.22).

Remark 4.18. Mismatches between the driving and response system have been lumped into a nonlinear function $\Theta(z,v,u)$ which has been interpreted as an augmented state variable η to obtain system (4.35). In this way, η represents the uncertain functions $\alpha(z,v)$ and $\beta(z,v)$ which include the transmitter/receiver discrepancy. Then, if one is able to stabilize the system (4.35) around the origin via output feedback, then the trajectories of system (4.22) will converge to a neighbourhood containing the origin. Finally, an estimated value $\beta_E(z)$ can be computed from Lie derivative of the output function $h(x)$ along the vector field $g(x_S)$.

Theorem 4.19. Systems (4.32) and (4.33) are practically synchronized if the equilibrium point $(z^*,\eta^*,v^*) = (0,0,0)$ of system (4.22) is semiglobally practically stabilizable.

Lemma 4.20. Consider the following state-feedback: $u = -(\eta + K^T z)/\beta_E(z)$, where k_i's, $i = 1,2,\ldots,\rho$, are coefficients of the Hurwitz polynomial $P_\rho(s) = s^\rho + k_1 s^{\rho-1} + \ldots + k_{\rho-1}s + k_\rho$. Under state feedback the solution $(z(t),\eta(t),v(t))$ converges to the origin if the initial conditions satisfy $\Psi_0 = 0$.

Proof. Suppose that the initial conditions satisfy $\Psi_0 = 0$. Thus since $\Psi(z,\eta,v,u)$ is an invariant manifold, $\eta = \Theta(z,v,u)$ for all $t \geq 0$. Then, we have that $\eta = \alpha(z,v) + \delta(z,v)u$

where $\delta(z,v) = \beta(z,v) - \beta_E(z)$. Combining η with state feedback controller we have that $u = -(\alpha(z,v) + K^T z)/\beta(z,v)$. Since $\Delta f(x,\pi)$, $g_M(x_M)$ and $g_S(x_S)$ are smooth vector fields, $\alpha(z,v)$ is a smooth function of its arguments. Consequently, the augmented state $\eta(t)$ and the state feedback controller $u = u(z)$ are bounded. On the other hand, convergence of the states $(z(t),\eta(t))$ follows from the fact that the subsystem (4.35) is in cascade form and the corresponding characteristic polynomial is Hurwitz under state feedback controller. In addition, if $(z(t),\eta(t)) = 0$ and the subsystem $= \zeta(0,v)$ converges to the origin (minimum phase assumption), the state feedback is a practical stabilizer of system (4.35).

Remark 4.21. Note that controller in Lemma 4.20 requires full information about the state of system (4.35). In this sense following comments are in order: (i) Lemma 4.20 implies that closed loop system (i.e., the discrepancy system under control actions) can be written as $\dot{z} = Az$ where the matrix $A \in \mathbb{R}^{n\times n}$ is Hurwitz; that is it has all its eigenvalues at the open left-hand complex plane, (ii) the augmented state η is not available for feedback, this fact is obvious because η represents, by definition, the mismatches between drive and response systems, (iii) it is desired that only one state is transmitted in chaos-based communication schemes, then only one state $x_i = x_{i,M} - x_{i,S}$ is available for feedback from *on-line* measurements. Consequently, estimated values of states (z,η) are required for practical implementation. According to recent results estimated values of partial feedback can be obtained from robust estimation procedures, see for instance [45].

Remark 4.22. The state feedback $u = u(z(t),\eta(t))$ requires a priori knowledge about the augmented state $\eta(t)$. According to Assumption A.2, that is an unrealistic situation. Hence, an estimated value of the uncertainties is desired. We are interested in a dynamic output feedback of the form

$$\begin{aligned}\dot{\hat{z}}_i &= \hat{z}_{i+1} + L^i\kappa_i(z_1 - \hat{z}_1); \qquad i = 1,2,\ldots,\rho-1\\ \dot{\hat{z}}_\rho &= \hat{\eta} + \beta_E(\hat{z})u + L^\rho\kappa_\rho(z_1 - \hat{z}_1)\\ \dot{\hat{\eta}} &= L^{\rho+1}\kappa_{\rho+1}(z_1 - \hat{z}_1)\end{aligned} \tag{4.37}$$

$$u = Sat\left\{\frac{1}{\beta_E(\hat{z})}[\hat{\eta} + K^T\hat{z}]\right\} \tag{4.38}$$

where $Sat\{\cdot\}:\mathbb{R}^n \to S \subset \mathbb{R}^n$, S is a bounded set and $(\hat{z},\hat{\eta})$ are estimated values of (z,η), respectively. The κ_j's, $j = 1,2,\ldots,\rho+1$, and k_i's, $i = 1,2,\ldots,\rho$, are chosen in such way that they are coefficients of Hurwitz polynomials and $L > 0$ is the high-gain estimation parameter. The dynamic output feedback (4.37) is able to suppress chaos with uncertain vector fields [25] and thus, in principle, it can be used to unmask signals via chaotic synchronization.

Lemma 4.23. Under actions of the dynamic output feedback (4.38), the origin of the nonlinear system (4.35) is semiglobally practically stabilizable.

Proof. Let $e \in \mathbb{R}^{\rho+1}$ be a vector whose components are defined by: $e_j := L^{\rho+1-j}(z_1 - \hat{z}_1)$, $j = 1,2,\ldots,\rho$, $e_{\rho+1} := \eta - \hat{\eta}$. In this way the closed-loop system is obtained substituting controller (4.37) and dynamics of above defined estimation error into equation (4.35). Thus closed loop system becomes

$$\begin{aligned} \dot{z} &= Z(z,\eta,e;L) \\ \dot{\eta} &= \Gamma(z,\eta,e,u,\dot{u}) \end{aligned} \tag{4.39}$$

$$\begin{aligned} \dot{v} &= \zeta(z,v) \\ \dot{e} &= LA'(\kappa;r)e + \varphi_2(z,\eta,e;L)B' \end{aligned} \tag{4.40}$$

where $z \in \mathbb{R}^{\rho}$, $v \in \mathbb{R}^{n-\rho}$, $Z(z,\eta,e;L) = Az + [\eta + \beta_E(\hat{z})u]B$, where B is the ρ-dimensional vector $[0,\ldots,0,1]^T$, the nonlinear $\beta_E(\hat{z})$ is given as above and $A \in \mathbb{R}^{\rho\times\rho}$ is a companion matrix given by

$$A = \begin{bmatrix} 0 & 1 & 0 & \cdots & 0 \\ 0 & 0 & 1 & \cdots & 0 \\ \vdots & \ddots & 0 & \ddots & \vdots \\ 0 & \ddots & \ddots & \ddots & 1 \\ 0 & 0 & \cdots & 0 & 0 \end{bmatrix}$$

The feedback function takes the form $u = u((z - N(L)e_\rho),(\eta - e_{\rho+1}))$, where $e_\rho = [e_1,\ldots,e_\rho]^T$, $N(L) = \mathrm{diag}(L^{-\rho},L^{-\rho+1},\ldots,L^{-1})$ and $\Gamma(z,v,\eta,u,u)$ is given as above. Note that since the saturation function is bounded, there exists a continuous function $\gamma(|e|)$ such that $|Z(z,\eta,e;L) - Z(z,\eta,0;L)| < \gamma(|e|)$. In addition note that, since $\eta = \alpha(z,v) + \delta(z,v)u$ and $u = \mathit{Sat}\{(\hat{\eta} - K^T\hat{z})/\beta_E(\hat{z})\}$, one can obtain the contraction $\eta = F(z,\eta,u,e;L)$ (which can be computed from first integral of system (4.35), *i.e.*, $\eta = \int\Gamma(z,v,\eta,u,u)d\sigma$). Then, according with the Contraction Mapping Theorem, the state η can be expressed globally and uniquely as a function of the coordinates (z,e). On the other hand, the matrix $\acute{A}(\kappa) \in \mathbb{R}^{(\rho+1)\times(\rho+1)}$ is the companion matrix of the estimator (4.38) and is given by

$$A'(\kappa;r) = \begin{bmatrix} -\kappa_1 & 1 & 0 & \cdots & 0 \\ -\kappa_2 & 0 & 1 & \cdots & 0 \\ \vdots & 0 & 0 & \ddots & \vdots \\ -\kappa_\rho & \vdots & \ddots & \ddots & 1 \\ -r(t)\kappa_{\rho+1} & 0 & \cdots & 0 & 0 \end{bmatrix}$$

$\acute{B}$ is a $(\rho+1)$-dimensional vector given by $[0,\ldots,0,1]^T$ and $r(t) = [\beta_E(\hat{z}) - \mathrm{Sat}'\{\cdot\}\delta(z,v)/\beta_E(z\hat{})]$ where $\mathrm{Sat}'\{\cdot\}$ is the time derivative of the saturation function. The nonlinear function $\varphi_2(z,\eta,e;L) = \Gamma(z,v,\eta,u,\dot{u}) - L\kappa_{\rho+1}e_1$ is a continuous and bounded function.

Now, since the polynomial $s^{\rho+1} + \kappa_1 s^{\rho} + \ldots + \kappa_{\rho-1}s + \kappa_{\rho+1} = 0$ is Hurwitz (see Remark 4.21), there exists any positive number, ε, satisfying $0 < \varepsilon < 1$ (see Theorem 2.7 in [45]). Hence the nominal system $\dot{e} = \acute{A}(\kappa;r(t))e$ is quadratically stable for $|1 - r(t)| < \varepsilon$. This implies that the Lyapunov equation $\acute{A}^T(\kappa;r(t))P + PA(\kappa;r(t)) = -I$ has a positive-definite solution. Since the nonlinear function $\varphi_2(z,\eta,e;L)$ is bounded, the subsystem (4.41) is quadratically asymptotically stable.

From this and the boundedness of $Z(z,\eta,e;L)$ and $\phi_2(z,\eta,e;L)$, we conclude that (applying Theorem 3 in [19]) given a compact set of initial conditions $X_0 \subset \mathbb{R}$ containing the origin, there exists an upper bound U^{MAX} with $|Sat\{\cdot\}| \leq U^{MAX}$ and a high-gain estimation parameter L such that X_0 is contained in the attraction basin $U_{PS} \times U_{PSo}$. Hence system (4.40 - 41) is semiglobally practically stable.

Proof of Theorem 4.19. Suppose that the nonlinear system (4.35) is semiglobally practically stable, that is $(z,v,\eta) \to 0$. Then, by minimum phase assumption and Theorem 4.17, i.e., solution of system (4.22) is a projection of system (4.35), one has that the states $(z,v) \to 0$ via module $\Pi\cdot(z,\eta,v)$. In addition, according to Lemma 4.6, the coordinate transformation $z = \phi(x)$ is globally defined. Consequently as $\Pi\cdot(z,\eta,v) \to 0$ as $(z,v) \to 0$ $\Rightarrow x(t) \to 0$. This implies that $x_S(t) \to x_M(t)$ for all $t \geq 0$.

Corollary 4.24. The message signal s_m is demodulated by the dynamic output feedback (4.38) if and only if systems (4.32) and (4.33) are completely practically synchronized.

Proof. (I) According to Theorem 4.17, there exists an upper bound U^{MAX} with $Sat\{\cdot\} \leq U^{MAX}$ and a high-gain estimation parameter L such that systems (4.32) and (4.33) are practically synchronized via the dynamic output feedback (4.38). Then from system (4.34) and Assumption (A.3) one gets that $u/s_m \approx \Delta f(x,\pi) g_M(x_M)/g_S(x_S)$. This means that $\|\Delta f(x,\pi)\| \approx 0$ (complete synchronization) $\Rightarrow \|u - s_m\| \approx 0$. (II) Suppose that $u = s_m$. Then from system (4.34) and Assumption (A.3) one has that at the equilibrium $x^*(t) = 0$, $\Delta f(x,\pi) g_M(x_M)/g_S(x_S) \approx u - s_m = 0$. Since $g_M(x_M)$ and $g_S(x_S)$ are bounded away from zero therefore above condition implies that $\Delta f(x,\pi) \approx 0$.

4.3.3 Illustrative Examples

We present two examples in this Section. First example consists in two third order system whose model is similar but parameter values are different. The aim is to show that message signal can be recovered in spite of parametric variations and illustrate that chaotic minimum phase assumption is satisfied. In addition, we have assumed that transmission channels are equal, *i.e.*, $g_S(x_S) = g_M(x_M)$. This assumption was considered in order to isolate parametric differences. Second example consists in two strictly different oscillators. Here the goal is to show that message signal can be acceptably recovered in spite of model differences between transmitter and receiver. We choose two second order driven oscillators to illustrate this case. Thus drive system is given by Duffing equation whereas response system is given by van der Pol oscillator. In second example we have assumed that transmission channels are unequal, *i.e.*, $g_S(x_S) \neq g_M(x_M)$.

Signal Transmission in spite of parametric variations. Suppose that modulated and demodulated signal is carried out by same channel, *i.e.*, $g_S(x_S) = g_M(x_M) = (1,0,0)^T$. Chua circuit has been chosen to illustrate the proposed decoding scheme. This oscillator is widely studied as a carrier for secure communications schemes. The drive system can be written in dimensionless form as follows

$$\begin{aligned}\dot{x}_{1,M} &= \pi_{1,M}[x_{2,M} - x_{1,M} - f(x_{1,M})] + s_m \\ \dot{x}_{2,M} &= x_{1,M} - x_{2,M} + x_{3,M} \\ \dot{x}_{3,M} &= -\pi_{2,M}x_{2,M}\end{aligned} \tag{4.41}$$

where $f(x_{1,M}) = \pi_{3,M}x_{1,M} + ½(\pi_{4,M} - \pi_{3,M})(|x_{1,M} + 1| - |x_{1,M} - 1|)$. Suppose that the same configuration is used as a response system. However, assume that there are differences between the electronic devices (for instance, the nonlinear resistance). That is, the parameter values of the receiver are different than the transmitter. In this way the response system becomes

$$\begin{aligned}\dot{x}_{1,S} &= \pi_{1,S}[x_{2,S} - x_{1,S} - f(x_{1,S})] + u \\ \dot{x}_{2,S} &= x_{1,S} - x_{2,S} + x_{3,S} \\ \dot{x}_{3,S} &= -\pi_{2,S}x_{2,S}\end{aligned} \tag{4.42}$$

where $f(x_{1,S}) = \pi_{3,S}x_{1,S} + ½(\pi_{4,S} - \pi_{3,S})(|x_{1,S} + 1| - |x_{1,S} - 1|)$, with $\pi_S \neq \pi_M$.

From the differences $x_i = x_{i,M} - x_{i,S}$, the discrepancy system (4.22) can be obtained as follows: $\dot{x}_1 = \Delta f_1 + s_m - u$, $\dot{x}_2 = \Delta f_2$, $\dot{x}_3 = \Delta f_3$. Now defining the transmitted state by $y_M = x_{1,M}$ and the receiver output by $y_S = x_{1,S}$, one has that $h(x) = x_1$. This implies that the smallest integer such that: (i) $L_g L_f^i h(x) = 0$, $i = 1,2,\ldots, \rho\text{-}2$ and (ii) $L_g L_f^{\rho} h(x) \neq 0$ is $\rho = 1$. In this way the coordinate transformation is globally defined and becomes: $z_1 = x_1$, $v_1 = x_2$ and $v_2 = x_3$. Then the discrepancy system can be rewritten as

$$\begin{aligned}\dot{z}_1 &= \Delta f_1 + s_m - u \\ \dot{v}_1 &= \Delta f_2 \\ \dot{v}_2 &= \Delta f_3 \\ y &= z_1\end{aligned} \tag{4.43}$$

where Δf_i are unknown functions and y denotes the output of the discrepancy system.

In order to illustrate that system (4.44) satisfies the minimum phase assumption, one can show that $\Delta f_2 = z_1 - v_2 + v_3$ and $\Delta f_3 = -2\gamma_2 x_2 + \delta_2$, where $\gamma_2 = ½(\pi_{2,M} + \pi_{2,S})$ and $\delta_2 = \pi_{2,M}x_{2,S} + \pi_{2,S}x_{2,M}$ are uncertain; however it is clear that $\gamma_2 > 0$ and δ_2 is bounded. As $z_1 \to 0$ (zero dynamics), one has that: $= Cv + \Delta_2$, where $\Delta_2 = [0\ \delta_2]^T$ and

$$C = \begin{bmatrix} -1 & 1 \\ -2\gamma_2 & 0 \end{bmatrix}$$

Hence since $\gamma_2 > 0$ and Δ_2 is bounded, the zero dynamics is asymptotically stable. That is, discrepancy between systems (4.42) and (4.43) is a minimum phase system.

Since Assumptions (A.1)-(A.4) are satisfied, the augmented state can be defined as: $\eta = \Delta f_1 + s_m$. Then, system (4.35) can be constructed and the system for the dynamic output feedback (4.37-38) is given by

$$\begin{aligned}\dot{\hat{z}}_1 &= \hat{\eta} - u + L\kappa_1(z_1 - \hat{z}_1) \\ \dot{\hat{\eta}} &= L^2\kappa_2(z_1 - \hat{z}_1) \\ u &= Sat\{\hat{\eta} + k_1\hat{z}_1\}\end{aligned} \tag{4.44}$$

where

$$Sat\{u\} = \begin{cases} U^{MAX} & if \quad u > U^{MAX} \\ \hat{\eta} + k_1 \hat{z}_1 & if \quad -U^{MAX} \leq u \leq U^{MAX} \\ -U^{MAX} & if \quad u < -U^{MAX} \end{cases}$$

where $U^{MAX} = 7.5$ was arbitrarily chosen. The control gain was chosen to be $k_1 = 7.0$. The estimation constants, κ_1 and κ_2, were chosen in such way that the polynomial $s^2 + \kappa_1 s + \kappa_2 = 0$ has its roots at -1.0. The high gain parameter L = 100.0. The initial conditions were arbitrarily chosen as follows: $x_M(0) = (0.1,-0.5,0.5)$, $x_S(0) = (0.6,0.4,-0.8)$ and $(\hat{z}_1(0),\hat{\eta}(0)) = (0.1,-0.2)$.

Figure 4.10 shows the synchronization of the drive and response system. Here there is no message signal ($s_m = 0$). The feedback was turned on at t = 30 s. Note that the dynamic output feedback (4.45) yields complete synchronization. Indeed, equal drive and response model was used in this numerical simulation. Consequently, the trajectories of the discrepancy system converges exactly to the origin (CES).

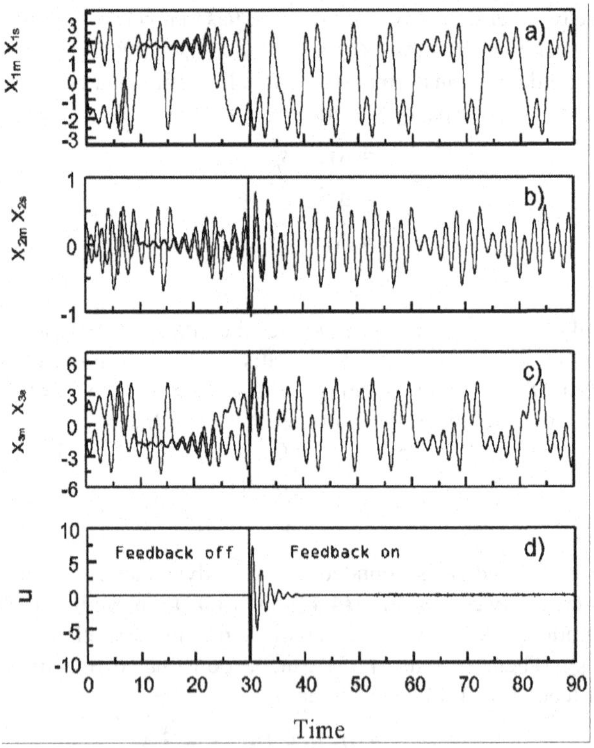

Fig. 4.10. Complete synchronization via the dynamic output feedback (4.45). Drive system state are tracked by the response system.

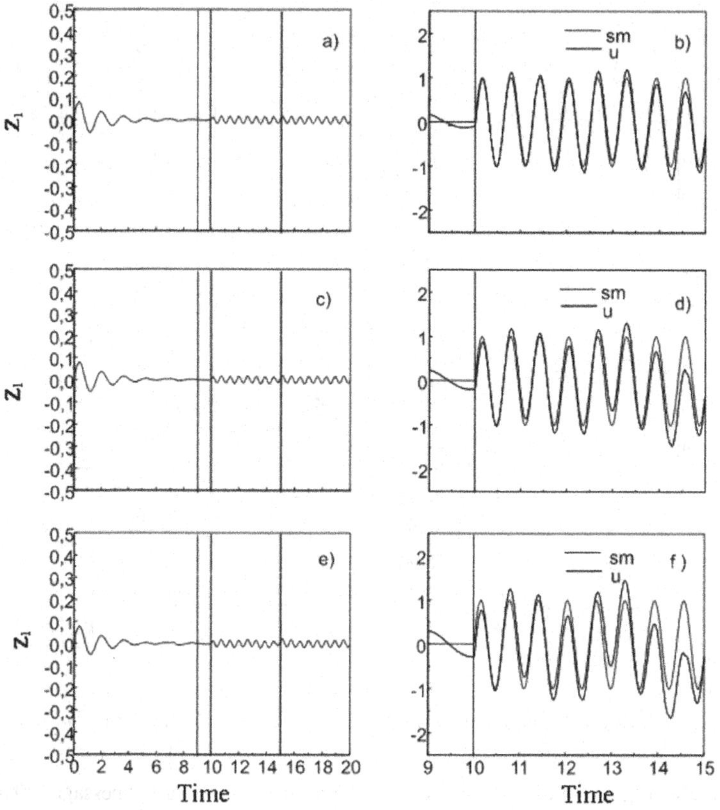

Fig. 4.11. Transmission of an analog signal. The output feedback (4.37-39) unmasks the message signal in spite of transmitter/receiver mismatches. The response system parameters were modified by, (a) 5%, (b) 10% and (c) 15%.

Figure 4.11 shows the performance of the proposed scheme. Here the dynamic output feedback was turned on at t = 0 and the message signal was transmitted for $t \geq 10$ s. The message signal was chosen to be a periodic function $s_m = \sin(\omega t)$ where $\omega = 10$ rad/s.

The frequency was chosen such that the dynamic behavior of the drive system remains chaotic. The code/decoding scheme displays certain robustness margin against parametric variations. The parameters of the response system were varied 5% in Figures 4.11a and 4.11b while in Figures 4.11c and 4.11d, a variation of 10 % was induced. The message signal is decoded with acceptable accuracy in both cases. In fact, even if parameters of the response system are varied at 15 %, the message signal can be acceptably decoded (see Figures 4.11e and 4.11f).

Note that since mismatches between driving and response system are present, the trajectories of the discrepancy system converge to an arbitrarily small neighborhood containing the origin (complete practical synchronization). The same performance

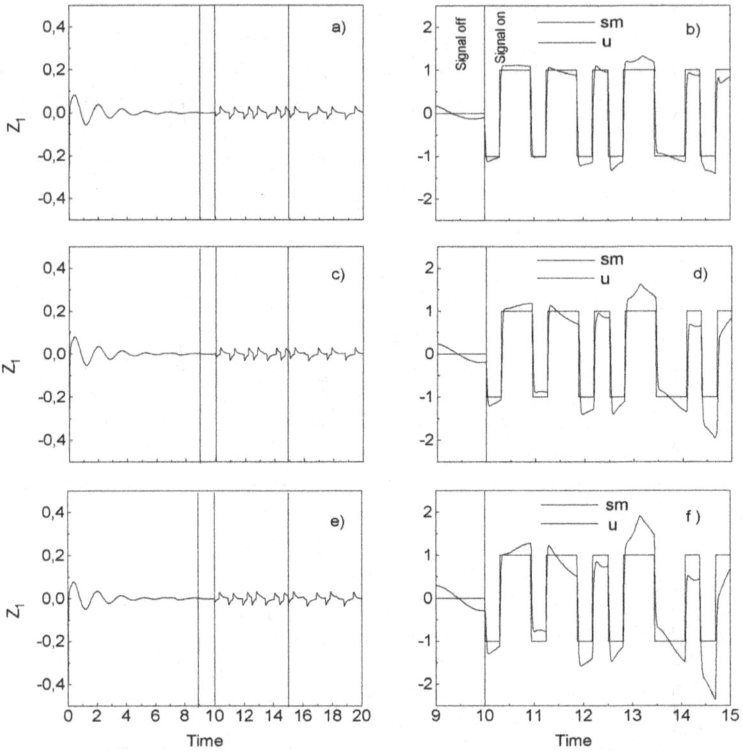

Fig. 4.12. The digital signal transmitted was chosen as the word "message". The output feedback (6) recovers the message signal in spite of the transmitter/receiver mismatches. The response system parameters were modified by: (a) 5%; (b) 10%; (c) 15%.

was found when the message signal consists of a digital word. The transmitted signal consisted of the digital word "message". Figure 4.12 shows the results for several parameters mismatches. The message signal was transmitted at 5 rad/s. Once again, the drive and response systems are completely practically synchronized and the message signal is acceptably recovered.

In order to obtain a measure of the performance we consider a performance index defined by $I := T\int_0^T |s_m(\sigma) - u(\sigma)|d\sigma$, where T is a constant. Figure 4.13 shows the index for several parameter values of the controller (4.45). The message signal was chosen in such way that the transmitter behaves chaotically, hence $s_m = 0.2\sin(10t)$. Initial conditions of drive system were chosen as $x_M(0) = (0.1,-0.5,0.5)$ while initial conditions of response system were chosen as (0.2,-0.3,-0.1). The controller (4.46) was activated at t = 0 and the estimation constants were chosen as $\kappa_1 = 2$ and $\kappa_1 = 1$. The message signal was entered at t = 0. Here $\%\Delta\alpha := \pi_{1,S} = \pi_{1,M}(1 + \Delta\pi_{1,M})$ and $\%\Delta\beta := \pi_{3,S} = \pi_{3,M}(1 + \Delta\pi_{3,M})$. Thus, for example, $\%\Delta\alpha = -10$ means that $\pi_{1,S} = 0.9\pi_{1,M}$ and $\%\Delta\beta = 5$ means that $\pi_{3,S} = 1.05\pi_{3,M}$. The intervals of $\%\Delta\alpha$ and $\%\Delta\beta$ were chosen in such way that systems (4.42) and (4.43) behaves chaotically. Note that as $\%\Delta\alpha \rightarrow 0$

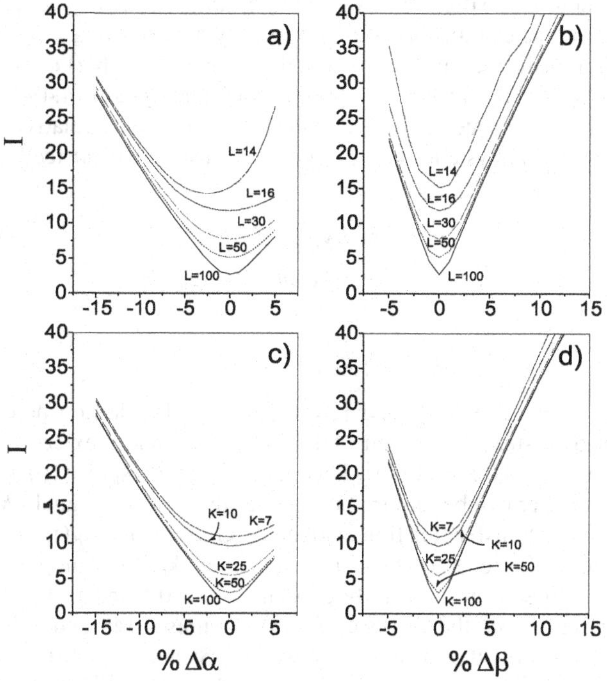

Fig. 4.13. Performance index versus parametric differences of drive and response systems. An analog signal was transmitted. Note that as the differences tends to zero as the recovering is better.

and $\%\Delta\beta \rightarrow 0$ as the index I decreases. This implies that message signal is recovered with acceptable accuracy.

Signal transmission in spite of strictly different model. The goal of the example is to illustrate the message signal can be recovered: (i) in spite of different model of drive and response systems which is the extreme case of transmitter/receiver mismatches, (ii) external perturbations by oscillatory signals which can be interpreted as noise and (ii) transmission channels are different, *i.e.*, $g_S(x_S) \neq g_M(x_M)$. Thus, we chose Duffing equation as transmitter and van der Pol oscillator as receiver. Drive system becomes

$$\dot{x}_M = f_M(x_M;\pi_M) + g_M(x_M)s_m + \tau_M(t) \tag{4.45}$$

where $f_M(x_M) = (x_{2,M}, x_{1,M} + x31_{,M} - 0.15x_{2,M})^T$, $g_M(x_M) = (0, 2)^T$, $\tau_M(t) = (0, 0.275\cos(1.1t))^T$ and the message signal is given by $s_m = 0.25\sin(2t)$. Response system is given by

$$\dot{x}_S = f_S(x_S;\pi_S) + g_S(x_S)u + \tau_S(t) \tag{4.46}$$

where $f_S(x_S) = (x_{2,s}, 0.01(1 - x_{1,S})x_{2,S} - x_{1,S})^T$, $g_M(x_M) = (0, 1)^T$, $\tau_M(t) = (0, 1.273\cos(0.3t))^T$. If $h_1(x) = x_{1,M}$ and $h_2(x) = x_{1,S}$ are the comparison states, and defining $x_i = x_{i,M} - x_{i,S}$, $i = 1,2$, one gets the following discrepancy system

$$\begin{aligned} \dot{x} &= \Delta f(x;\pi) + g_M(x_M)s_m - g_S(x_S)u + \Delta\tau(t) \\ y &= x_1 \end{aligned} \tag{4.47}$$

where y is the output of the discrepancy system, $\Delta f(x;\pi) = f_M(x_M) - f_S(x_S)$, $\Delta\tau(t) = \tau_M(t) - \tau_S(t)$. Thus the coordinates transformation is given by $z_1 = x_1$ and $z_2 = x_2$. In such a way system (4.48) is transformed into $z_1 = z_2$ and $z_2 = \alpha(z,t;\pi) + \beta(z)u$, where $\alpha(z,t;\pi) = \Delta f(z;\pi) + g_M(x_M)s_m + \Delta\tau(t)$ and $\beta(z) = g_S(x_S)$. Note that system (4.48) is FLNS, *i.e.*, there is no unobservable states v in discrepancy system because relative degree $\rho = 2$. Now, defining $\eta := \alpha(z)$ system (4.35-36) can be constructed and feedback (4.37-38) becomes

$$
\begin{aligned}
\dot{\hat{z}}_1 &= \hat{z}_2 + L\kappa_1(z_1 - \hat{z}_1) \\
\dot{\hat{z}}_2 &= \hat{\eta} + g_S(x_S)u + L^2\kappa_2(z_1 - \hat{z}_1) \\
\dot{\hat{\eta}} &= L^3\kappa_3(z_1 - \hat{z}_1) \\
u &= -\hat{\eta} + k_1\hat{z}_1 + k_2\hat{z}_2
\end{aligned}
\qquad (4.48)
$$

where $L = 5$, $\kappa = (3.0, 3.0, 1.0)^T$ and $K = (25.0, 5.0)$. In addition note that if exact synchronization of system (4.46) and (4.47) is attained under external perturbations $\Delta f(x;\pi) = 0$ and $x_M - x_S \equiv 0$ for all $t \geq 0$. This implies that $u = s_m(g_M(x_M) + \Delta\tau(t))/g_S(x_S)$. That means that a filter can be required to recover the message signal. We designed a low pass filter whose transfer function is given by $G(s) = u_r/u = k_r/(\tau_r s + 1)$ where u_r is the filtered signal, τ_r is characteristic time of filter, k_r is an amplification factor. Parameters of the filter were arbitrarily chosen as $k_r = 0.5$ and $\tau_r = 3.9$. Figure 4.14 shows the performance of the feedback (4.49). Figures 4.14a and 4.14b shows the dynamical evolution of the discrepancy states. Systems (4.46) and (4.47) are synchronized. Moreover, message signal is acceptably recovered with delay which is induced by filter (see Figure 4.14c).

This scheme consists in a dynamic output feedback which performs the suppression of chaos on the discrepancy system. The dynamic output feedback is designed by means of the following algorithm: (i) uncertainties into the discrepancy system are lumped in a nonlinear function and (ii) the lumping nonlinear function is interpreted as an augmented state. In particular, we have chosen a dynamic output feedback based on high-gain estimation and linearizing-like controller. However, the scheme is not restricted to this class of controller. Indeed, adaptive control or robust control theory can be used to design a dynamic output feedback which should render the chaos suppression of the augmented system of the discrepancy system.

In addition, the scheme allows recovering message signals in spite of parametric variations and strictly different model. Since the drive and response systems are nonlinear, the transmitter/receiver mismatches can yield additive (or multiplicative) uncertainties. This class of uncertainties can be studied via H_∞ control theory [46].

Therefore, we have chosen a dynamic output feedback based on high-gain estimation and linearizing-like controller. However, the proposed scheme is not restricted to this class of controller. Indeed, adaptive control or robust control theory can be used to design a dynamic output feedback which should render the chaos suppression of the augmented system of the discrepancy system. In addition, the scheme allows recovering message signals in spite of the parametric variations and strictly different models.

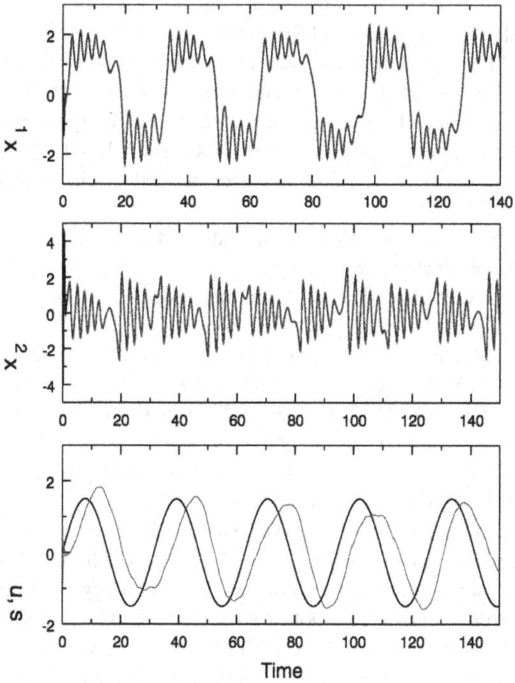

Fig. 4.14. Recovering a message signal when drive system model is strictly different to response system model. Simulations were carried out in presence of external perturbing signals. A low pass filter was used and message signal is acceptably recovered.

Exercise 4.25. *Hyperchaotic synchronization.* Consider two hyperchaotic systems given by the Rössler systems [47] $x_1 = -(x_2 + x_3)$, $x_2 = x_1 + ax_2 + x_4$, $x_3 = 3 + x_3x_1$, $x_4 = -0.5x_3 + 0.005x_4$ and consider the parameter $a_M = 0.25$ and $a_S = 0.28$. Perform the procedure described in this section to completely synchronize both systems.

Exercise 4.26. *Synchronization of two strictly different chaotic systems.* Consider the Lorenz system as the master and the Rössler third order system as the slave. Perform the procedure in order to obtain synchronization and obtain the function that relates the slave states with those of the master (generalized synchronization [48]).

References

[1] Pecora, L.M., Carrol, T.S.: Synchronization in chaotic systems. Phys. Rev. Lett. 64, 821 (1990)

[2] Wu, C.W., Yang, T., Chua, L.O.: On adaptive synchronization and control of nonlinear dynamical systems. Int. J. of Bifur. and Chaos 6(3), 455 (1996)

[3] Kapitaniak, T., Sekeita, M., Ogorzalek, M.: Monotone synchronization of chaos. Int. J. of Bifur. and Chaos 6(1), 211 (1996)

[4] Wu, C.W., Chua, L.O.: Synchronization in an array of linearly coupled dynamical systems. IEEE Trans. Circuits and Systems I 42, 430 (1995)
[5] di Bernardo, M.: An adaptive approach to the control and synchronization of continuous-time chaotic systems. Int. J. of Bifur. and Chaos 6(3), 557 (1996)
[6] Parlitz, U., Kocarev, L.: Using Surrogate Data Analysis for Unmasking Chaotic Communication Systems. Int. J. of Bifur. and Chaos 7(2), 407 (1996)
[7] Perez, G., Cerdeira, H.: Extracting messages masked by chaos. Phys. Rev. Lett. 74, 1970 (1995)
[8] Collins, J.J., Stewart, I.N.: Hexapodal gaits and coupled nonlinear oscillators model. Biological Cybernetics 68, 287 (1993)
[9] Collins, J.J., Stewart, I.N.: Coupled nonlinear oscillators and the symmetries of animal gaits. J. of Nonlinear Science 3, 349–392 (1993)
[10] Kozlov, A.K., Shalfeev, V.D., Chua, L.O.: Exact synchronization of mismatched chaotic systems. Int. J. of Bifurc. and Chaos 6, 569 (1996)
[11] Mossayebi, F., Qammar, H.K., Hartley, T.T.: Adaptive estimation and synchronization of chaotic systems. Phys. Lett. A 161, 255 (1991)
[12] Chua, L.O., Yang, T., Zhong, G.C., Wu, C.W.: Adaptive synchornization of Chua oscillator. Int. J. of Bifur. and Chaos 6, 189 (1996)
[13] Woafo, P., Chedjou, J.C., Fostin, H.B.: Dynamics of a system consisting of a van der Pol oscillator coupled to a Duffing oscillator. Phys. Rev. E 54(6), 5929–5934 (1996)
[14] Femat, R., Alvarez-Ramírez, J., González, J.: A strategy to control chaos in nonlinear driven oscillators with least prior knowledge. Phys. Lett. A 224, 271 (1997)
[15] Femat, R., Alvarez-Ramírez, J., Castillo-Toledo, B., González, J.: On robust chaos suppression in a class of nondriven oscillators: Application to Chua's Circuit. IEEE Circuits and Syst. I 46, 1150 (1999)
[16] Nijmeijer, H., van der Shaft, A.J.: Nonlinear Dyanamical Systems. Springer, Berlin (1990)
[17] Isidori, A.: Nonlinear control systems. Springer, NY (1989)
[18] Herman, R., Krener, A.J.: Nonlinear controllability and observability. IEEE Trans. on Automatic Control AC-22, 728 (1977)
[19] Esfandiari, F., Khalil, H.K.: Output feedback stabilization of fully linearizable systems. Int. J. of Control 56, 1007 (1992)
[20] Sussman, H.J., Kokotovic, P.V.: The peaking phenomenon and the global stabilization of nonlinear systems. IEEE Trans. on Automatic Control AC-36, 424 (1991)
[21] Bazhenov, M., Huerta, R., Rabinovich, M.I., Sejnowski, T.: Cooperative behavior of a chain of synaptically coupled chaotic neurons. Physica D 116, 392 (1998)
[22] Wu, C.W., Chua, L.O.: A simple way to synchronize chaotic systems with application to chaotic secure communications. Int. J. of Bifur. and Chaos 3, 1619 (1993)
[23] Short, K.M.: Steps toward unmasking secure communication. Int. J. of Bifur. and Chaos 4, 959 (1994)
[24] Niels, H., Holstein, R., Kay, P.Y., Sosnovttseva, O.V., Mosekilde, E.: Synchronization phenomena in nephron-nephron interactions. Chaos 11, 417 (2001)
[25] Giugliano, M., Bove, M., Grattarola, M.: Insulin release at the molecular level: Metabolic-electrophysiological modeling of the pancreatic beta-cells. IEEE Trans. on Biom. Engineering 47, 611 (2000)
[26] Femat, R., Jauregui-Ortiz, R., Solís-Perales, G.: A chaos-based communication scheme via robust asymptotic feedback. IEEE Trans. on Circuits and Systems I 48, 1161 (2001)
[27] Mettin, R., Lauterborn, W., Hubler, A.: A Scheeline, Parametric entrainment control of chaotic system. Phys. Rev. E 51, 4065 (1995)

[28] Kokarev, L., Parlitz, U., Hu, B.: Lie derivatives and dynamical systems. Chaos, Solitons and Fractals 9, 1359 (1998)
[29] Aeyels, D., Sepulchre, R.: Stability for dynamical systems with first integrals: a topological criterion. Syst. & Contr. Lett. 19, 461 (1992)
[30] Ushida, A., Nishio, Y.: Spatio-temporal chaos in simple coupled chaotic circuits. IEEE Trans. On Circ. and Syst. I 42, 678 (1995)
[31] Shalfeev, V.D., Osipov, G.V.: The evolution of spatio-temporal disorder in a chain of unidirectionally-coupled Chua's circuits. IEEE Trans. on Circ. and Syst. I 42, 687 (1995)
[32] Huerta, R., Bazhenov, M., Rabinovich, M.I.: Clusters of synchronization and bistability in lattices of chaotic neurons. Europhysics Letters 43, 719 (1998)
[33] Cuomo, K.M., Oppenheim, A.V.: Circuit implementation of synchronization chaos with applications to communications. Phys. Rev. Lett. 71, 65 (1993)
[34] Kocarev, J.L., Parlitz, U.: General approach for chaotic synchronization with application sto communication. Phys. Rev. Lett. 74, 5028 (1995)
[35] Grassi, G., Mascolo, S.: A system theory approach for designing cryptosystems based on hyperchaos. IEEE Trans. Circuits and Systems I 46, 1135 (1999)
[36] Nijmeijer, H., Mareels, M.Y.: An observer looks at synchronization. IEEE Trans. Circuits and Systems I 44, 882 (1997)
[37] Liao, T.-L., Huang, N.-S.: An observer-based approach for chaotic synchronization with application to secure communication. IEEE Trans. Circuits and Systems I 46, 1144 (1999)
[38] Takens, F.: Detecting strange attractors in turbulence. In: Rand, D.A., Young, L.S. (eds.). Lectures Notes in Mathematics, vol. 898, p. 361. Springer, Berlin (1980)
[39] Ott, E., Grebogi, C., Yorke, J.A.: Controlling chaos. Phys. Rev. Lett. 64, 1196 (1990)
[40] Chen, G., Dong, X.: From chaos to order: Perspectives and methodologies in controlling chaotic nonlinear dynamical systems. Int. J. of Bifurc. and Chaos 3, 1363 (1993)
[41] Jackson, E.A.: Control of dynamics flow with attractors. Phys. Rev. A 44, 4839 (1991)
[42] Ogorzalek, M.J.: Taming chaos-Part I: Control. IEEE Trans. Circuits and Systems I 40, 700 (1993)
[43] Teel, A., Praly, L.: Tools for semiglobal stabilization by partial state and output feedback. SIAM J. of Contr. Opt. 33, 1443 (1991)
[44] Femat, R., Solís-Perales, G.: On the chaos synchronization phenomena. Phys. Lett. A 262, 50 (1999)
[45] Khargonekar, P.P., Patersen, I.R., Zhou, K.: Robust stabilization of uncertain linear systems: Quadratic stabilizability and H_∞control theory. IEEE Trans. Automatic Control AC-19, 356 (1990)
[46] Zhou, K., Doyle, J.C.: Essentials of Robust Control. Prentice-Hall, NJ (1998)
[47] Rössler, O.E.: An equation for hyperchaos. Phys. Lett. A 71, 155 (1979)
[48] Rulkov, N.F., Sushchik, M.M., Tsimring, L.S., Abarbanel, H.D.I.: Generalized synchronization of chaos in directionally coupled chaotic systems. Phys. Rev. E 51, 980 (1995)

5 Discrete-Time Feedback for Chaos Control and Synchronization

5.1 Discrete - Time Control of Systems with Friction

Now a discrete-time approach to feedback controller is discussed to control a particular system which describes a friction phenomenon. This system is used to introduce some features of the discrete time controller for chaos control. The friction system comprises some interesting dynamical properties, *e.g.*, an invariant manifold characterized by zero velocity and velocity direction. Such properties allow a practical justification of the feedback design, which yields control of the measured state and its time derivative. The mechanical justification and the feedback design allow us to introduce the synergetic interpretation. Here the *self*-organization of a simple dynamical system is discussed from the understanding of the effect of control parameters acting over mechanical systems. The control parameter is yielded by the so-called *controller*. The controller is a feedback scheme from a finite-differences approximation. Such justification leads us to develop a chaos suppression scheme. The main idea is to counteract the nonlinear forces acting onto (or into) the systems and compensates the external perturbation forces acting over the nonlinear systems. The goal is to compute an estimate value of the uncertain force in such way that nonlinear systems can be controlled. This is, the synergetics of the second-order driven oscillators is studied from the point of view of the control theory. In principle, the finite-difference is able to achieve chaos control and synchronization. In addition, we shall see that a discrete time approach feedback attains synchronization against master/slave mismatches. Indeed, the procedure yields synchronization of strictly different oscillators. In this sense, it is said that controller is *robust*. This means that *self-organization* of this class of oscillators can be achieved in spite of master/slave mismatches (even if oscillators are strictly different). We belief that synergetics is due to feedback structure *into* the nonlinear system.

A discrete-time feedback controller is designed for a class of nonlinear engineering system. The system consists in two metallic plates. The upper plate is moving on the

R. Femat & G. Solis-Perales: Robust Syn. of Chaotic Sys. Via Feedback, LNCIS 378, pp. 139–175, 2008.
springerlink.com

lower, which is fixed. The system has friction between plates, hence its dynamical model is given by a second-order driven oscillator [1]. The system is nonlinear due to the friction force. One can assume that the friction model is not known. This assumption is reasonable, since the friction model is not easy to obtain (see for instance [1], [2] and [3]). Indeed, this is theme of actual research [4]. This assumption implies that the friction system can be interpreted as an uncertain nonlinear model. Since the friction system contains uncertainties, a dynamical estimated value is provided by discrete-time feedback. In this sense, the feedback has adaptive structure. This is, an internal model based on the fundamental structure of the friction system is constructed. Thus, an estimated value of the internal model of the mechanical system can be provided by a dynamic estimator. However, the controller does not require a reference model of the system; *i.e.*, the internal model is neither adapted nor adjusted. In fact, the controller only requires (a) Measurements of the position and (b) Knowledge about the last control command. The uncertainties about the friction model and external perturbations are lumped into a nonlinear term, which is estimated from finite differences scheme. After that, the feedback counteracts the nonlinear forces and external perturbations. Indeed, in some sense, such feedback interconnections can induce self-organization.

5.1.1 Position Regulation of a Friction System

Let us consider the model of a mechanical system with friction given by [1]:

$$m\ddot{x} + F(x,\dot{x}) + \alpha x = \tau_1(t) + u \tag{5.1}$$

where m represents the mass of the upper plate system, $F(x,\dot{x})$ includes all the friction terms, α is a system parameter, $\tau_1(t)$ is an unknown time function, which may be due to loads and/or noise acting on the mechanism, and u is a manipulated force used to controlling the system via feedback. Inaccuracies in mechanical system are often caused by the presence of friction. Typical errors caused by friction steady-state errors and tracking lags in position trajectories. The former are mainly caused by static (or dry) friction, which is proportional to the velocity direction. The latter are generated by viscous friction. Previous results shown that to deal with friction, it is necessary to have a good characterization of the structure of the friction model and design appropriate compensation schemes [1]. The impact of friction on the performance of precision control system has received some attention: a number of works have recently appeared that discuss the modelling and compensation of friction. Among proposed concepts for dealing with friction is to estimate its force and generate a control to counteract it [1], [2], [3].

A lot of effort has been devoted to the modelling of friction. It is well established that friction depends on the direction of the moment. Phenomena such as sticking friction (torque needed to start the motion), and a downward bend at low velocities have been identified [4]. Detailed experiments performed at low velocities have confirmed the Tustin's model, which includes a decaying exponential term. In general, it can be said that friction is a complicated phenomenon representing all the forces opposing the motion. Such forces can be both orthogonal and tangential to the direction of motion. It is easy to conclude that friction forces present in mechanical systems are hard to be modelled and inherently unknown. In addition, the friction

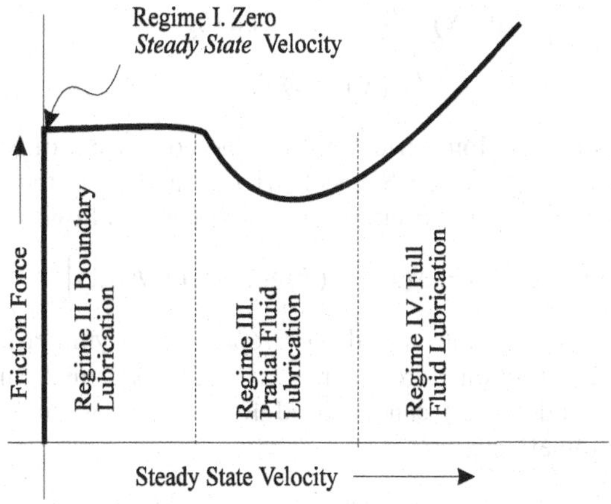

Fig. 5.1. Friction regimes as a function of the stationary velocity

system can be interpreted as the forces interaction, which are dynamically coupled via interconnections. In this sense, the friction control can provide a preliminary result toward self-organization.

5.1.2 The Friction Force and Its Complexity

Friction models have been extensively studied [1],[2] and [4]. It is well established that friction forces are function of the velocity. Although there is disagreement on the character of the functionality of the friction forces with the velocity, experiments have confirmed that, for moderate and low velocities, the main components in the friction forces are mainly caused by the following phenomena [4] (see Figure 5.1):

a) *Coulomb and sticktion.* Coulomb friction is due to sticking effects. There is a constant friction torque opposing the motion when the velocity is not zero. For zero velocity, the stiction oppose all motions as long as the forces are smaller in magnitude than stiction force.

b) *Stribeck (downward bends).* After the sticktion force has been surmounted, the friction force decreases exponentially, reaching a minimum, and then increases proportionally with the velocity. These bends occur at velocities close to zero. The friction forces are due to a partial lubrication, where the velocity is adequate to entrain some fluid in the junction but not enough to fully separate the surfaces.

c) *Viscous.* These forces appear at nonzero velocity due to energy dissipation in the lubricant fluid contained between the moving surfaces. Here the surfaces are fully separated by fluid film.

d) *Asymmetries and position dependence.* Imperfections and unbalances in the mechanism induce asymmetries and position dependence of the friction forces. However, experiments on industrial mechanism have shown that this dependence is relatively weak [1],[4].

The following expressions can be used to model the friction effects (a)-(c)

$$F_C(\dot{x}) = \beta_1 \operatorname{sgn}(\dot{x}) \tag{5.2a}$$

$$F_S(\dot{x}) = \beta_2 \exp(-\mu|\dot{x}|)\,\mathrm{sgn}(\dot{x}) \quad (5.2b)$$

$$F_V(\dot{x}) = \beta_3 \dot{x} \quad (5.2c)$$

where β_1 represents the Coulomb friction, β_2 is the coefficient of Stribeck friction, μ represents the slip constant in the Stribeck friction and β_3 is the coefficient of the Viscous friction. In this way, the friction can be written as follows:

$$F(x,\dot{x}) = \varphi(x)\left[F_C(\dot{x}) + F_S(\dot{x}) + F_V(\dot{x})\right] \quad (5.3)$$

The function $\varphi(x)$ is introduced to represent asymmetries and position dependencies of the friction forces or a normal load that may change with displacement. The model (5.3) can be used for simulating (at least for moderate velocities) real friction effects.

5.1.3 The Irregular Behavior of the Friction System

Note that, due to the function sign($\dot{x}$), Friction system is discontinuous at $\dot{x} = 0$. A consequence of the discontinuity in sign($\dot{x}$) is the stick-slip motion [4]. When the trajectory passes through $\dot{x} = 0$, the static friction may balance the external forces. When this happens, the system remains stuck at zero velocity until the driven forces reaches the value of the Coulomb friction β_1.

Geometrically, sticking behavior is present when some regions in the phase portrait where the flow of the equation (5.3) is directed from x > 0 toward the set $\Sigma = \{\dot{x} = 0\}$, and simultaneously the flow of the directions. These regions of conflict are the sticking region $R \subset \Sigma$. Formally, it is possible to define an equivalent vector field on R which is (n-2)-dimensional (zero velocity and consequently, constant position) if the original vector field is n-dimensional. From a control theory viewpoint, this phenomenon is responsible for the existence of steady-state offsets in position. On the other hand, from the synergetics point of view, such region can be seen as a consequence of the interaction between the friction phenomena. This is, self-organization of the involved vector fields result in a *conflict* region given by a geometrical constraint. As a consequence of the reduction of the dimensionality of the flow generated by (5.3), the dynamical behavior of the friction system is irreversible (*i.e.*, it is possible to recover the past of the trajectory by only changing the time sign. In other words, the phase portrait of (5.3) is not invariant under the transformation $t \rightarrow -t$). The consequence of the discontinuity of the vector field in (5.3) is that embedding of an observable variable is not globally diffeomorphic to the phase flow, so that control of the system (5.3) cannot be attained via delay coordinates techniques [5].

To show that the system (5.1) can display complicated behaviors, let us consider the parameter $\beta_1 = 1.0$, $\beta_2 = \beta_3 = 0.0$, the functionality

$$\varphi = \begin{cases} 1 + kx & \text{for} \quad x > -\dfrac{1}{k} \\ 0 & \text{for} \quad x \le -\dfrac{1}{k} \end{cases} \quad (5.4)$$

and the external forces $\tau_1(t) = A \sin(\Omega t)$ and $u = 0$.The function $\varphi(x)$ represents a normal load which vary with the position. Note that $\varphi(x)$ is continuous; this is, the unique source of discontinuities is $\mathrm{sign}(\dot{x})$.

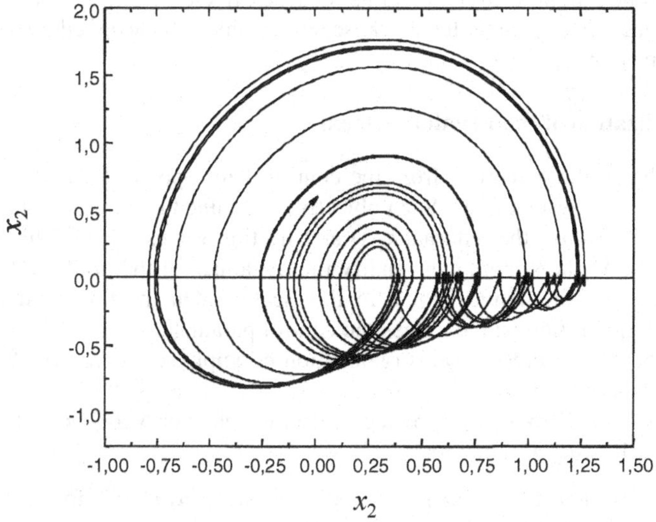

Fig. 5.2. Phase portrait of the system (5.1) without control actions

Figure 5.2 shows the $(x, \dot{x})$-portrait for the case $\Omega = 1.25$, $A = 1.9$ and $k = 1.5$. One can interpret the Figure 5.2 as being the phase portrait of the following nonlinear second-order driven oscillator system $\ddot{x} + F(x,\dot{x}) + \alpha x = 0$, which can be perturbed by the signal u. In the unforced phase portrait, one can distinguish two regions R and S in the set Σ (with $R \cup S = \Sigma$). In the set $S = \{\dot{x} = 0, x \leq \delta^*\}$ for certain $\delta^* < 0$, the direction both vectors agree, such that the function sign(x) suffers only one switching. On other hand, in the set $R = \{\dot{x} = 0, x > \delta^*\}$ the flow of the system is in conflict so that R is a sticking region. It is said that $R=\{\dot{x} = 0\}$ is a set of weak equilibrium points because a small perturbation *slides* slowly the trajectories toward the strong equilibrium point (0,0). When the flow of the system is perturbed with an harmonic force $\tau_1(t) = A \sin(\Omega t)$, almost all the points of the set S remain as switching points. This invariance is due to the *transversality* of the trajectories at R. On other hand, the set R induces a band B of *sliding* flow (small velocity dynamics with alternating stiction periods), where the trajectories move to the regions with lower normal load $\varphi(x)$. The trajectory leaves the *sliding* band B in a region where the force $\tau_1(t)$ becomes larger than the Coulomb friction to return again to the sliding band B.

Complicated behavior, such as the one remarked above, can be controlled in most cases because degradation of the mechanical parts may undesirable. Thus, some kind of control actions are necessary in order to induce more regular behavior despite of friction and external forces. This is an engineering reason for the use of a feedback to control the position of the upper plate. However, from the point of view of the nonlinear science, it is important to understand the effect of the feedback action onto dynamical systems. This is a realistic situation. The main idea behind this control study is to compute a feedback force such that the synergism of vector fields can be modified. Hence, a new dynamical order (which could be regular or chaotic) is displayed by the nonlinear system and its feedback interconnections. In what follows, we present the discrete-time feedback scheme, which is designed to regulate the position of the upper plate.

5.1.4 Stabilization of the Friction System

On the one hand, the objective, from the control theory point of view, is to stabilize the behavior of the system (5.1). Such objective is attained by means of a manipulated force that compensates the external $\tau_1(t)$ and friction forces $F(x,\dot{x})$. This is, coupled forces *acting onto* the system can modify the dynamics of the nonlinear systems. In fact, we study an alternative choice in this section. On the other hand, from the synergetics, the question is: How does the control parameter affect to the dynamics of the system (5.1)?. Is there any self-organization principle under feedback?. This point is discussed in last Section.

The control problem is to generate a desire behavior via feedback, u. Let us consider the following assumptions:

(A.1.1) Position $x = x_1$ and velocity $\dot{x} = x_2$ are available for feedback from measurements.

(A.1.2) The parameter α and the plate mass m are available for feedback whereas the friction forces $F(x,\dot{x})$ are not available for feedback.

(A.1.3) External perturbing force, $\tau_1(t)$, is bounded for all $t \geq 0$.

Some comments regarding the assumptions are in order. (A.1.1) is physically realizable though shaft encoder measurements. Except for robots, in most mechanical systems the mass m is fixed, so that it can be accurately known. On other hand, α can be interpreted as a string constant, which, in most cases, is also well known. Assumption (A.1.2) is a realistic situation for practical applications. Assumption (A.1.3) is not strong. The physical perturbations are often bounded.

Let us write the system (5.1) as a set of first-order equations

$$\begin{aligned} \dot{x}_1 &= x_2 \\ \dot{x}_2 &= -\alpha x - F(x_1,x_2) + \tau_1(t) + u \end{aligned} \tag{5.5}$$

where $(x_1,x_2) = (x,\dot{x})$. Without lost of generality, we have assumed that $m = 1$. Let $r(t) \in \mathbb{C}^2$ be a desired position trajectory, for instance, a fixed position. If the friction forces were known, the following ideal feedback

$$u^T(x,t) = -\tau_1(t) + \alpha x_1 + F(x_1,x_2) + V(r) \tag{5.6}$$

where $V(r) = \ddot{r} + g_1(x_2 - \dot{r}) + g_2(x_1 - r)$ drives asymptotically the system trajectories to a behavior with $x_1 = r(t)$. The control parameters g_1, g_2 are chosen in such a way that the matrix

$$M=\begin{bmatrix} 0 & 1 \\ -g_1 & -g_2 \end{bmatrix}$$

has all its eigenvalues at the open left side of the complex plane. The control feedback (5.6) will be called to as "*Ideal Feedback Control*" (IFC) because, in order to control, it requires perfect knowledge about system. The main advantage of controller (5.6) is that the controlled system is linear. When the friction forces $F(x_1,x_2)$ are not known, an option is to construct an approach of the feedback. Nevertheless, according to Assumption (A.1.3.), the friction forces are not available for feedback. One basic idea is to estimate this uncertain terms.

Let us consider the be a lumping term defined by $\eta(t) = F(x_1(t),x_2(t)) + \tau_1(t)$. Then, the system (5.5) can be written as follows

$$\dot{x}_1 = x_2 \tag{5.7a}$$

$$\dot{x}_2 = -\alpha x_1 + \eta(t) + u \tag{5.7b}$$

In principle, at each time $t \geq 0$, the uncertain term $\eta(t)$ can be estimated by solving the acceleration, $\dot{x}_2 = \ddot{x}$, from the equation (5.7a) as follows

$$\eta(t) = \dot{x}_2 + \alpha x_1 + \tau_2 \tag{5.8}$$

This is, the friction plus exogenous forces can be estimated via a torque balance. In order to synthesize a feedback (which can suppress the erratic behavior displayed if the Figure 5.2) with a estimation like (5.8), we have to use approximations for the acceleration, $\dot{x}_2 = \ddot{x}$. Let $\Delta > 0$ the measuring rate and consider the control law in the interval $t \in [t_k, t_{k+1})$, $t_{k+1} - t_k = \Delta$. Then one has that

$$u_k = -\eta_k + \alpha x_{1,k} + V(r_k) \tag{5.9a}$$

$$V_k(r) = \ddot{r}_k + g_1(x_{2,k} - \dot{r}_k) + g_2(x_{1,k} - r_k) \tag{5.9b}$$

$$\hat{\eta}_k = \frac{(x_{2,k} - x_{2,k-1})}{\Delta} + \alpha x_{1,k} - u_k \tag{5.10}$$

where, at $t = t_k$, $x_{2,k}$ is the measured, $\dot{x}_2$ is the acceleration of the system, $\hat{\eta}_k = \hat{\eta}(t_k)$ represents the estimated value of the uncertain term $\eta(t_k)$ and $u_k = u(t_k)$ is the feedback command, which is used to control the system. Note that as $\Delta \to 0$, the estimate value of the uncertain term given by the equation (5.10) converges to $\eta(t_k)$ in (5.8), then $u_k = u^I(t_k)$, and from (5.9) and (5.10), the following discrete-time feedback can be obtained $u(x(t_k)) = \dot{x}_{2,k} + u(t_{k-1}) + V_k(r)$; from where $\dot{x}_2 = V_k(r)$. Consequently, in the limit as $\Delta \to 0$ the behavior under the control (5.9), (5.10) converges to the behavior under the IFC (5.6). The control (5.9) can be interpreted as a Δ-perturbation of the IFC (5.6). Evidently, the case $\Delta = 0$ can not be implemented because in the limit $\Delta \to 0$ the control (5.9), (5.10) becomes ill-defined.

The estimated value of uncertain forces, η_k, has the following interpretation. The combination of (5.9) and (5.10) leads to the expression of the uncertain term: $\hat{\eta}_k = \dot{x}_{2,K} - V_k(r) + \hat{\eta}_{k-1}$. On the other hand, introducing (5.9) into (5.7), we have the velocity dynamics for $t \in [t_{k-1}, t_k)$: $\dot{x}_{2,k} = V_k(r) + \hat{\eta}_{k-1}$. So therefore, $\hat{\eta}_k = (\eta_k - \hat{\eta}_{k-1})/2$. From above equality, one can see that estimated value is the arithmetic average of the actual friction forces $\eta_k = \eta(t_k)$ and last estimate $\hat{\eta}_{k-1}$. As $\Delta \to 0$, $\hat{\eta}_{k-1} \to \eta_k$. Hence, as lim $\Delta \to 0$ as $\hat{\eta} \to \eta(t)$. On the other hand, the average can be interpreted as a contraction mapping, with decaying rate equal to 0.5. If the friction forces $\eta(t)$ are bounded, the estimator (5.10) yields bounded estimate $\hat{\eta}_{k-1}$. So, the estimation scheme is stable. In fact, it has been shown that the system (5.1), under the feedback actions of the controller (5.9),(5.10), can track a dynamic reference signal [6].

The stability properties of the discrete-time feedback (5.9),(5.10) can be sketched as follows: Let us consider the system (5.1) under the control (5.9),(5.10) to get the control error $\dot{\varepsilon} = M\varepsilon + N(t)$, where $\varepsilon = (\varepsilon_1,\varepsilon_2) \in \mathbb{R}^2$ and $\varepsilon_1 = x_1 - r$, $\varepsilon_2 = x_2 - r$ and $N(t) = [0,\eta-\hat{\eta}]^T$. It is not hard to see that asymptotic error $\|\varepsilon(t\to\infty)\|$ is proportional to the error estimation $|\eta-\hat{\eta}|$. Consequently, the better estimation of η, the lower the control error $\|\varepsilon\|$.

Example 5.1. (*Position stabilization in a friction system*). Let us consider the system (5.5) with friction term given by function (5.3), where $\varphi(x)$ is given by equation (1.4). Assume that $\beta_1 = 1.0$ and $\beta_2 = \beta_3 = 0$, $\alpha = 1.0$ and $\tau_1 = 1.9\cos(1.25t)$. Now, assume that the controller (5.9),(5.10) is activated for $t \geq 30$ (*i.e.*, $u = 0$ for all $t < 30$). The control gain where chosen as $g_1 = 2.0$ and $g_2 = 1.0$, which implies that the polynomial $P_2(s) = s^2 + 2s + 1$ has all its roots located at -1.0. The sampling rate is $\Delta = 0.01$ and the reference signal is $r = 0$ (stabilization at origin).

Figure 5.3 shows the stabilization of the friction system at origin. Note that the time series of the measured state converges to a small neighborhood containing the origin (practical stability). However, the reader can verify that the IFC leads trajectories to the origin. In addition, notice that if control parameter is computed from (5.9),(5.10), high-frequency signals are displayed for certain intervals (see Figure 5.3c).

Unveiling feature is an additional property of the discrete-time feedback (5.9),(5.10). Although unveiling is not the goal of the paper, we present an example to illustrate that simple feedback laws reconstructs the forces acting into second-order oscillators. The aim is to show that, at time $t = t_k$, $k = 0,1,2,\ldots,\infty$, the feedback interconnections can yield a reconstructed value of the forces into the system. Figure 5.4 shows that the estimated value, $\hat{\eta}$, converges to actual values of the uncertainties, $\eta \equiv \tau_1(t) - F(x, \dot{x})$. The parameters were chosen as Figure 5.3.

Note that estimated value of the unavailable terms are provided by the feedback (5.9),(5.10). Then, it could be used to unveil unknown forces in second-order driven oscillators. This is, let us assume that a deterministic nonlinear system is, in some sense, self-organized. Besides, let us suppose that the nonlinear contains uncertain and unknown functions. In such case, the discrete-time feedback is able to unveil the uncertain function. To this end, the control command and the discrete-time estimator should be coupled. The procedure can be sketched as follows: The measured variable from the experimental apparatus can be input to a computer after filtering. The basic dynamical model (which can be called plant) and the feedback (5.9),(5.10) are

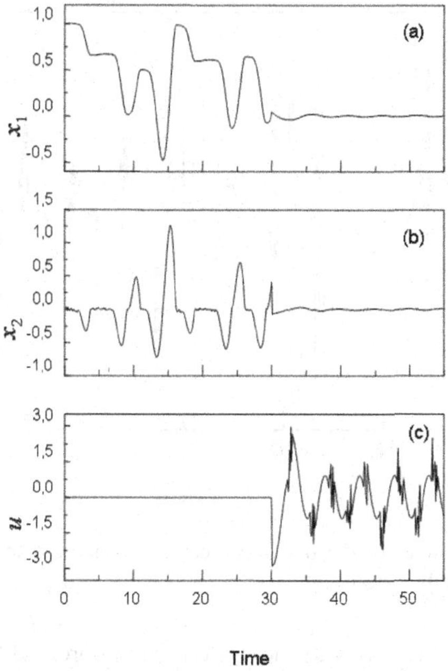

Fig. 5.3. Practical stabilization of the friction system (5.5). (a) Position, (b) velocity and (c) control action computed from equations (5.9), (5.10).

previously designed. The feedback is activated at t = 0. The control gains are chosen such that the polynomial $P_2(s)$ has its roots at the left-half complex plane.

Example 5.2. *Unveiling uncertain force acting into the system.* Consider a magnetic bearing whose dynamical model is given by $\ddot{x} + \delta\dot{x} + x + \alpha(x, \dot{x}, t) = \tau_1 + u$, where x means the position, δ is a damping factor, $\tau_1(t) = A\cos(\Omega t)$ represents a periodic perturbing force, $\alpha(x, \dot{x}, t)$ is a nonlinear function which represents the force between a high temperature superconductor and a magnet (which is supported by the Type-II superconductor). In [7], the authors reported that the characteristics of the system can depend on the hypo-elasticity function whose dynamics is given by $\dot{\alpha}(x,\dot{x}) = \mu_1[\alpha(x,\dot{x},t) - \varphi(x,\dot{x})]$ where μ_1 is a parameter and the nonlinear function is given by $\varphi(x,\dot{x}) = \varphi_1(x)[1 + \varphi_2(x)]$. Besides, the nonlinear functions can be approximated by $\varphi_1(x) = \mu_2\exp(-x)$ and

$$\phi_2(\dot{x}) = \begin{cases} -\mu_3 - \dot{x} & \text{for } \varsigma \le \dot{x} \\ \dfrac{-\dot{x}(\mu_3 + \mu_4)}{2\varsigma} & \text{for } -\varsigma \le \dot{x} < \varsigma \\ \mu_4 & \text{for } \dot{x} < -\varsigma \end{cases}$$

However, the nonlinear term $\alpha(x,\dot{x},t)$ is not exactly known. In addition, it has been shown that the superconducting magnetic bearing system given by above equation can display chaos [7]. This implies that it is highly sensitive to initial conditions. The

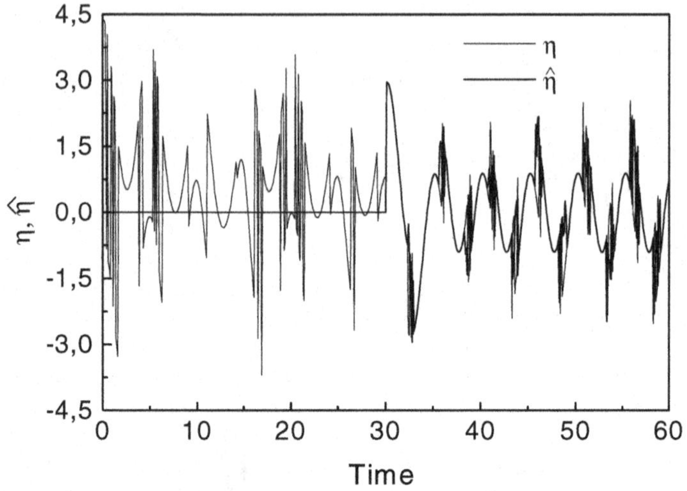

Fig. 5.4. The estimated values of the uncertain terms allows the stabilization of the nonlinear dynamical system (5.1) at the origin

questions are: Can the dynamics of the system be suppressed by means of the finite differences feedback?. (Consider the parameters values as follows: $\mu_1 = \mu_3 = \mu_4 = 1\ .0$, $\mu_2 = 0.3$, $\varsigma = 0.005$). In absence of time lags, can the feedback (5.9),(5.10) provide estimated values of the uncertain term, $\alpha(x,\dot{x},t)$?.

As illustrative example, let us assume that the unique uncertain term is the force between magnet and high-temperature superconductor, $\alpha(x,\dot{x},t)$. Suppose that only the position is measured. Besides the damping coefficient and the external perturbing force, $\tau_1(t)$ are exactly known. In such case, following the above procedure, the augmented state is defined by $\eta = \alpha(x,\dot{x},t)$. Hence, the discrete-time becomes: $x_2 \approx \hat{x}_{2,k} = [x_{1,k} - x_{1,k-1}]/\Delta$, $\eta(t_k) \approx \hat{\eta}_k = [x_{1,k} + 2x_{1,k-1} - x_{1,k-2}] + \delta\, \hat{x}_{2,k} + x_{1,k}\ \tau_1(t_k) + u_{k-1}$, $u_k = \hat{\eta}_k + (\delta + k_2) + (1 + k_1)x_{1,k} - \tau_1(t_k)$. Figure 5.5 shows the results of the control position of the levitation system from the above feedback scheme.

In resume, the discrete-time feedback has the following features: (a) It does not require knowledge about the nonlinear terms acting on the system. (b) It enters a linear behavior to the driven oscillator. In fact, the designed feedback retains the main characteristics of the IFC. A very simple and practical system has been used to illustrate the necessity of the dynamic estimation on nonlinear dynamic systems. Nevertheless, the proposed feedback has been developed under the consideration that the velocity (second time derivative of the main variable state) is available from measurements. However, in most cases, it is desirable a feedback scheme which depends on one measured state. This is not restrictive (see Example 5.2); at least for the case of second order systems. For instance, suppose now that only the position x_1 is available for measurements. In this case we use stable finite-differences to estimate acceleration: $\dot{x}_{2,k} \approx (x_{1,k} - 2x_{1,k-1} + x_{1,k-2})/\Delta^2$. In addition, the velocity is estimated via $x_{2,k} \approx (x_{1,k} - x_{1,k-1})/\Delta$. This means that the uncertain nonlinear terms requires only

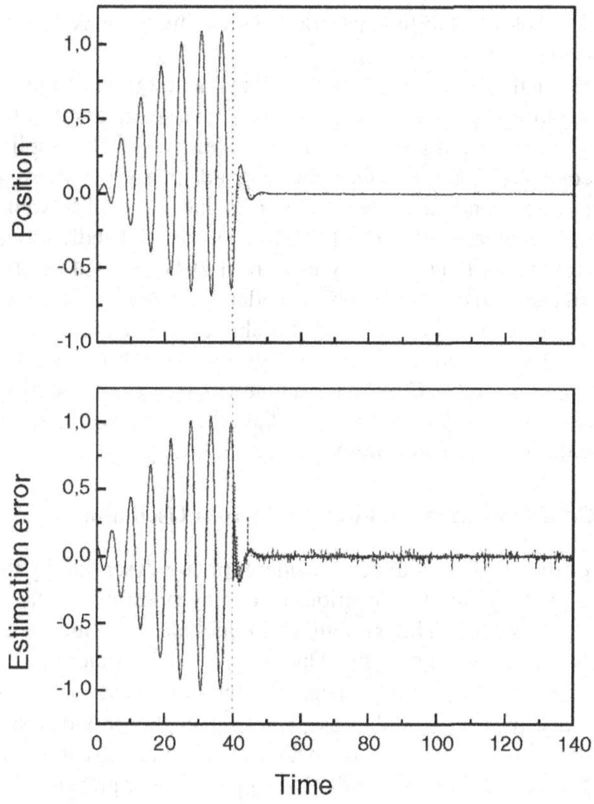

Fig. 5.5. Position control of the levitation system. The velocity was estimated from finite differences of the measured position.

position measurements. Indeed, the chaos suppression in second order driven oscillators can be achieved from the above procedure.

5.2 Chaos Suppression Via a Discrete-Time Controller

The control of chaotic systems has received increasing interest in last few years [5], [8], [9]. Ott, Grebogy and Yorke [5] provided the first strategy to stabilize periodic orbits embedded into a chaotic attractor. OGÝs strategy makes use of on-line construction of local invariant manifolds of the target orbit to derive a controller which counteracts the unstable directions of the orbit. Although OGY scheme is not robust against uncertainties and only leads to local stability, this control procedure possesses a feedback structure. This is, a contribution of the results in [5] is that feedback forces can suppress or induce chaos. More robust control strategies burrowed from conventional engineering methods of nonlinear control have been considered. These methods include linear control feedbacks [10] and adaptive control techniques [8]. A central issue when applying such feedback techniques is the obtainment of the robust stability margins against fluctuations in parameters (from external effects) and uncertain dynamic effects.

Linear control methods are based on a Taylor linearization of the system around a prescribed set-point. Although there exists a large amount of results about stabilization and trajectory tracking of linear systems [11], the application of linear techniques to control chaotic signals is not a good idea since chaos is of nonlinear nature. On the other hand, adaptive techniques have the disadvantage that linear structure in the parameters is assumed [12]. In addition, to handle model uncertainties with adaptive techniques is not an easy task. It must be pointed out that stabilization of uncertain systems is an exciting problem under intensive research [11].

The goal of this section is to extend the above discrete-time feedback from the friction system to the second-order chaotic systems, which include the Duffing, van der Pol, driven pendulum and Coulomb oscillators. As previous section, it is assumed that uncertainties in the model are present. That is, we address the robust stabilization problem of chaotic signals against model uncertainties.

5.2.1 Chaos Control of the Second-Order Driven Oscillators

The dynamics of the oscillators under consideration are described by the system (5.7). In general $F(x_1,x_2)$ is a nonlinear function depending possibly on the time to account for parametric excitations. The system (5.5) includes a wide variety of chaotic oscillators such as, for example, the Duffing and the Coulomb oscillator. In the former case $F(x_1,x_2) = x_1 - x_1^3 - \delta x_2$ whereas in the latter case $F(x_1,x_2)$ is given by the function of the dry friction, which was presented in the previous section [6], [13], [14]. In both cases, the external perturbation is periodic and it is given by $\tau_2(t) = \gamma\cos(\omega t)$. It is a well-known fact that the Duffing oscillator presents very complicated dynamics such as coexistence of chaotic attractors with periodic orbits [13]. On the other hand, the Coulomb oscillator displays erratic noninvertible trajectories due to a so-called stick-slip phenomenon (see Section 5.1).

In practical applications, it is desirable to induce regular dynamics in mechanical oscillators to avoid fracture and degradation of the mechanism parts. Persistent external perturbations represented by the time-functionalities in (5.5) are common in practice; for example, in the case of precise position mechanisms (*e.g.*, telescopes), where external vibrations and magnetic fields lead to errors and lag in position tracking. To avoid these undesirable dynamical effects, it is necessary to introduce some control actions in the system. Here we state our control problem: *given a target trajectory* $r(t)$, *design a control feedback* $\tau(x)$ *such that the system (5.5) tracks* $r(t)$. However, under light of the nonlinear science, such control problem implies that new organization is induced into the system by the feedback interconnection.

In seek of clarity, let us assume the following:

S.2.1) Only the position x_1 is available for measurements.
S.2.2) The nonlinear function $F(x_1,x_2)$ and the external force $\tau_2(t)$ are unknown.

Some comments regarding the above assumptions are in order. In most practical situations, it is only possible to measure position. Although in some case it is possible to get velocity measurements, it is an expensive and complicated procedure. On the other hand that $F(x_1,x_2)$ and $\tau_2(t)$ are not known is a realistic assumption since

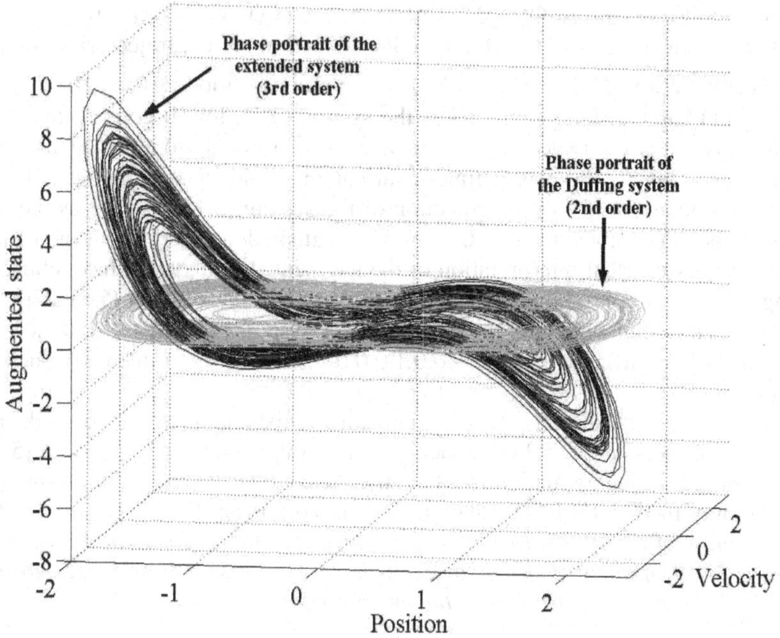

Fig. 5.6. Geometrical interpretation of the extended system via the uncertain term η(t)

imperfections of a mechanisms and unmodelled dynamics make impossible to obtain an accurate model of the oscillator. In the next section, a feedback control overcoming the above difficulties is provided. It should be noted that the above assumptions are less restrictive than made in previous section.

5.2.2 The Feedback Control without Velocity Measurements

Let $y(t) \in \mathbb{C}^2$ be a prescribed position trajectory (reference signal). If the terms $F(x_1,x_2)$ and $\tau_2(t)$ were known, then the controller given by equation (5.6) would steer asymptotically the trajectories of the system (5.5) to the desired reference. The parameters g_1 and g_2 are chosen in such a way that $P_2(\lambda) = \lambda^2 + g_1\lambda + g_2 = 0$ (which is the characteristic equation of the matrix M, see Section5.1) has its roots in the open left-half complex plane. But now, since the terms $F(x_1,x_2)$ and $\tau_2(t)$ are not known and the velocity is not available from measurements, an new alternative will be used to get estimated values of them; in such a way that the main characteristics of the IFC (5.6) are retained.

Once again, let us define $\eta(t) = F(x_1,x_2) + \tau_2(t)$. Thus, the system (5.5) can be rewritten as an extended space system in the following form

$$\begin{aligned} \dot{x}_1 &= x_2 \\ \dot{x}_2 &= \eta + u \\ \dot{\eta} &= \Gamma(x,\eta,\tau_1,t) \end{aligned} \tag{5.11}$$

where $\Gamma(x,\eta,\tau_1,t) = x_2\partial_1 F(x) + (\eta + \tau_1)\partial_2 F(x) + \dot{\tau}_2$ and $\partial_2 F(x) = \partial F(x)/\partial x_2$, $\partial_1 F(x) = \partial F(x)/\partial x_1$.

Notice that the manifold $\Phi(x) = \{(x,\eta) \in \mathbb{R}^3: \eta - F(x_1,x_2) - \tau_2(t) \equiv 0\}$ is invariant under the trajectories of (5.11). That is, $d\Phi(x)/dt = 0$ along the trajectories generated by the vector field in (5.11). Moreover, if the initial condition $(x(0),\eta(0)) \in \mathbb{R}^3$, the system (5.11) has the same solutions as the system (5.5), for all control inputs $\tau_1(t)$. Consequently, it is equivalent to design a controller for system (5.11) as for the system (5.5). This is, there is a time-invariant manifold, $\Psi(x,\eta,t;\pi)$, such that the solution of the system (5.5) is a projection of the system (5.11) as long as the initial conditions be $\Psi(x(0),\eta(0),0;\pi) \equiv 0$, which is satisfied by definition. In order to illustrate the geometrical interpretation of the augmented state, η, we have chosen the Duffing equation. Figure 5.6 shows the phase portrait of the systems (5.5) and (5.11). The initial conditions were chosen as follows: for system (5.5) $x(0) = (0.0,1.0)$ while for system (5.11) $(x_1(0),x_2(0),\eta(0)) = (0.0,1.0.0.01)$. The same parameters values were chosen for both systems.

Since systems (5.5) and (5.11) are dynamically equivalent, hence a feedback can be designed from the system (5.11) in such way that trajectories of the system (5.5) be leaded to origin. In fact, equation (5.11) is used as an intermediate system toward the construction of the feedback. The main idea is the following. *If one is able to stabilize the trajectories of the system (5.11) neither measurements of the velocity, x_2, nor the augmented state, η, then the trajectories of the system (5.5) will be leaded to origin (prescribed reference signal) against the uncertain terms.*

5.2.3 The Discrete-Time Feedback Via Uncertainties Estimation

A discrete-time controller can be designed for the system (5.11) by means of the procedure presented in Section 1. In this way, the IFC becomes:

$$\tau_2^I(x,t) = -\eta + V(y) \tag{5.12}$$

with $V(r)$ as in equation (5.6). In order to implement the controller (5.12), an estimated value of velocity x_2 and the unknown term η are required. In principle, x_2 and η can be calculated via position measurements as follows: $\dot{x}_2 = dx_1/dt$ and $\eta = d^2 x_1/dt^2 - \tau_1$ for all $t \geq 0$. Above equations state that the unknown variable η can be estimated via a torque balance (the second Newtońs law) while unmeasured velocity x_2 can be estimated by means of kinematic balance. This is an important property since it allows the design of feedback schemes based on nonlinear state estimators [9]. We provide a feedback based on time delay coordinates (TDC). The above estimation of velocity, x_2, and uncertainty, η, involve first and second-order derivations, respectively. Since it is not possible to find physical realization of derivations, an approximation of them must be used.

Assume that x_1 is measured at sampling frequencies Δ^{-1}, and consider the following feedback, which is held in the time interval $[t_k,t_{k+1}) = [t_k,t_k + \Delta)$,

$$\tau_{2,k} = -\hat{\eta} + V_k(y) \tag{5.13}$$

where

$$V_k(y) = \ddot{y}_k - g_1(\hat{x}_{2,k} - \dot{y}_k) - g_2(\hat{x}_{1,k} - y_k) \tag{5.14a}$$

$$\hat{\eta}_k = -\tau_{k-1} + \left(\hat{x}_{1,k} - 2\hat{x}_{1,k-1} + \hat{x}_{1,k-2}\right)/\Delta^2 \tag{5.14b}$$

$$\hat{x}_{2,k} = \left(\hat{x}_{1,k} - \hat{x}_{1,k-1}\right)/\Delta \tag{5.14c}$$

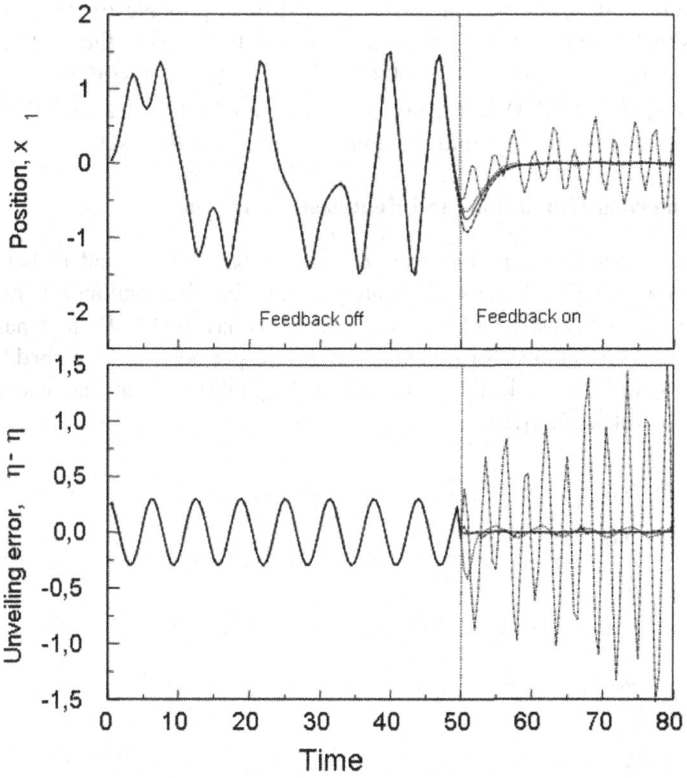

Fig. 5.7. Stabilization of the Duffing oscillator for several values fo the sampling rate D. (a) Position, (b) Control action.

As previous Section, subscript k designates values at time t_k. Backward finite-differences have been used in (5.14) to approximate velocity and acceleration at time t_k. Equation (5.14) contains an approximation to the torque balance to obtain an estimate of the unknown variable η and a kinematic balance to get an estimated value of the velocity. Now, let us define $\boldsymbol{x}_{1,k}^{(n)} = (x_{1,k}, x_{1,k}, \ldots, x_{1,k-n+1})$ be an n-dimensional TDC vector. The whole controller (5.13),(5.14) can be written as $\tau_{2,k} = \tau_2(\boldsymbol{x}_{1,k}^{(3)})$. This is, the control action at times $t \in [t_k, t_{k+1})$ depends only on the three-dimensional TDC vector $\boldsymbol{x}_{1,k}^{(3)}$. Taken's theorem [15] states that the dynamics of (5.1) can be embedded in a one-to-one way into a five-dimensional Euclidean space with coordinates vector $\boldsymbol{x}_{1,k}^{(5)}$. We have found here that the system (5.1) can be controlled with a control feedback depending only on a three-dimensional TDC vector [23].

For the stability proof, the following procedure can be sketched. Assume bounded trajectories of system (5.1). Backward finite-difference approximation in (5.14) yields $| a(t) - a_k | \leq \beta\Delta$, for certain positive number β. On the other hand, by using (5.14), one can show that $\eta(t) - \eta_k = a(t) - a_k$, $t \in [t_k, t_{k+1})$, from where $| \eta(t) - \eta_k | \leq \beta\Delta$. That is, the lower the sampling interval Δ, the lower the estimation error of the uncertainties $| \eta(t) - \eta_k |$. By using a continuous-time version of (5.13),(5.14), and with the aid of the stabilization results in [15], it is possible to state the converge properties of the controller scheme. One proves that as $\Delta \rightarrow 0$, the controller scheme leads to semiglobal asymptotic stability. This is, given a compact set of initial conditions $x_0 = x(0) \in D \subset \mathbb{R}^2$, one can find a controller such that $D \subseteq \Omega \subset \mathbb{R}^2$, where Ω is the region of asymptotic stability of the controlled system.

5.2.4 Numerical Simulations and Representative Examples

In order to illustrate the performance of the feedback (5.13) and (5.14), numerical simulations with the Duffing oscillator are provided. For this oscillator, $F(x,t) = x_1 - x_1^3 - 0.15x_2$ and $\tau_1(t) = 0.3\cos(t)$, which displays chaotic behavior [13]. From dynamic results, there is an infinite number of periodic and almost-periodic orbits embedded into the chaotic attractor [12]. Typically, such orbits and equilibrium points are chosen as target orbits for control actions [20].

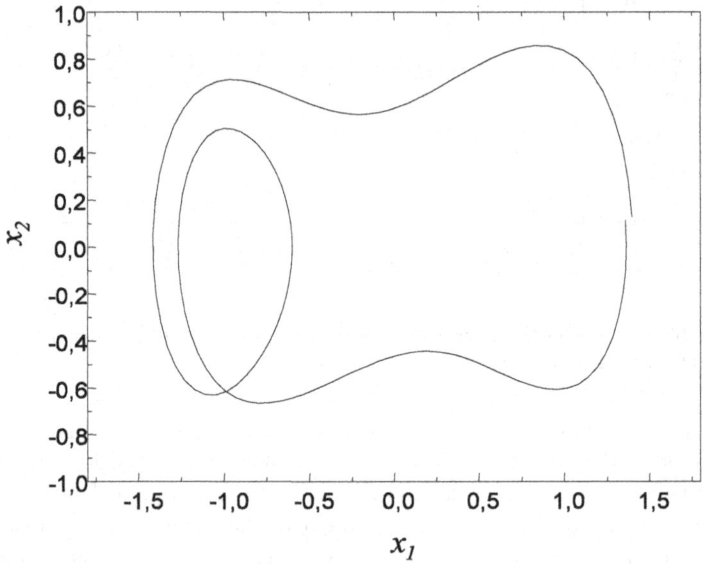

Fig. 5.8. Almost periodic orbit to be tracked

The origin (0,0) is an equilibrium point of the Duffing oscillator for $\tau_1(t) + \tau_2 = 0$. Such equilibrium is a saddle point. Without lost if generality, let us choose the constant signal as $r = 0$ as the control objective. The control parameters were chosen as $g_1 = 2$ and $g_2 = 1$,such that the roots of the polynomial $P_2(\lambda) = \lambda^2 + g_1\lambda + g_2$. are located at {-1,-1}. Here, $\Delta = 0.05$ was chosen, which is about 1/100 the period of the external force $\tau_1(t)$. Figures 5.7a shows the position trajectory before and after the control is activated at time t = 50 s. The trajectory is stabilized in the equilibrium point in spite of uncertainties in the model. Figure 5.7b shows the control signal. Stabilization of the origin is achieved when the control action τ_2 counteracts the external force $\tau_1(t)$ (*i.e.*, $\tau_2 \approx \tau_1(t)$).

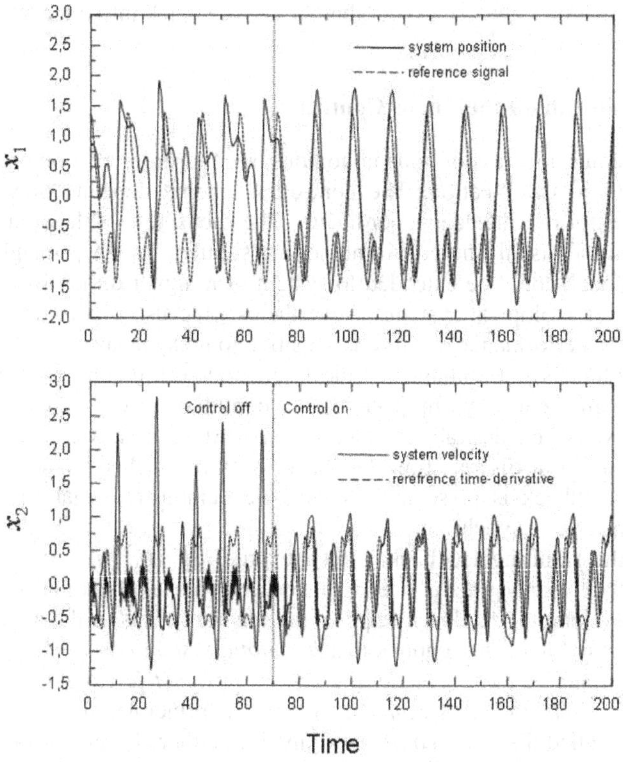

Fig. 5.9. Tracking of the almost-periodic orbit, which was taken from the Duffing oscillator

Let us now consider the trajectories tracking, *i.e.*, $r = r(t)$. Suppose that $r(t)$ is an almost-periodic orbit (which is shown in Figure 5.8). The almost-periodic orbits is a (finite-time) section of the chaotic trajectory, which was computed using the Duffing equation. We have selected a driven system where the nonlinear term is given by equation (5.2) (i.e., the friction system). Without lost of generality, the same parameter values than first Section were considered. Figure 5.9 displays the dynamical behavior of the controlled position $x_1(t)$, which after a transient, the Coulomb system tracks the almost-periodic behavior. In this case, the controller is activated at t = 75 s., which implies that for all $t \leq 75$ s. the control signal is $\tau_2 = 0$.

It should be noted that the designed feedback is able to track a desired (arbitrary) bounded signal. The feedback torque, τ_2, only requires on-line measurements of any observable state from where the estimated values of the uncertain terms (and unmeasured states) are computed. The feedback comprises two parts: a linear estimator and a linear feedback law. In addition, notice that the feedback induces a linear behavior from a linear estimator. Figure 5.9 suggests that, since tracking of dynamic reference signal can be attained, chaos synchronization can be achieved against master/slave mismatches. In particular, if feedback (5.11),(5.12) is interconnected, only one state and its time derivative are required. In this sense, self-organization of two unpredictable deterministic (and possibly chaotic) systems is carried out. Therefore, one can expect that simple feedback interconnections structure performs synergetics.

5.2.5 Discussion about the Chaos Control

A strategy to control a class of nonautonomous, second-order chaotic oscillators have been presented in this Section. The controller scheme depends only on a three-dimensional vector of time delay coordinates. The robustness of the controller against model uncertainties is illustrated via numerical simulations. In principle, the above presented feedback could be extended to system of a higher dimension; for instance, flow of fluids or biological systems. Nevertheless, the use of the finite differences could yield poor performance due to it is sensitive to noisy measurements. In fact, this discrete procedure is not suitable for the high order estimation, because it is very sensitive to high frequency. In principle, a continuous version of the controller (5.13),(5.14) would be desirable. However, in next section, we discuss a general procedure to design a discrete-time feedback. Moreover, chaos synchronization via the proposed feedback is presented. In fact, we demonstrate that strictly different chaotic systems can be synchronized.

Note that the resulting feedback is on the order of $n+1$, where n represents the order of the dynamical system. This is, in order to get and estimated value of the second order system, the feedback requires the measurements of the available state at the time t_k, t_{k-1} and t_{k-2}. This implies that *a chaotic system can be controlled from a (n+1)-order estimator, even if modelling errors, parametric variations and external perturbations affects to the dynamical system.* In principle, the above conjecture (which will be called the *order conjecture*) involves either continuous or discrete-time control systems. In what follows, such conjecture is studied. We belief that order conjecture is related to the *relative degree*. Actually, synchronization of high-order strictly-different systems is an open problem.

5.3 Discrete-Time Chaos Control and Its Implications

In this section, an approach to the discrete-time feedback (which was developed in the last Section) is performed. The goal of this section is to provide a procedure to design and analyze a discrete-time feedback (including the finite-differences algorithm). In addition, the control of nonlinear driven oscillators is posed is the context of the chaos control. In this way, the discrete-time chaos control is posed as a general theory. The aim of the Section is to provide evidence regarding the *order conjecture* for the case where the feedback is designed as a discrete-time controller.

Although chaos control was initially considered to be mostly of academic interest, recent applications have arisen in diverse disciplines (see [17], [18] and references therein). It has been pointed out that it is important not only to understand and describe chaotic motion but even to control dynamics into some desirable motion. Essential elements in the control of chaos are the following: (a) Stabilization of a chaotic system about a given reference and (b) suppression of 'erratic' dynamics. In this sense, we can characterize two main problems that arise when dealing with the control of chaotic systems: *Synchronization and suppression of Chaos.* Both synchronization and suppression are related, in some sense, to the synergetics notion.

The *synchronization problem* consists of making two or more chaotic systems behave in a synchronous way. This could seem impossible, if we think on the following features of the chaotic motion: (a) The sensitive dependence of initial conditions. In practice, it is almost impossible to reproduce the same starting conditions in master and slave. (b) To match exactly the model parameters of two chaotic systems. Even infinitesimal variations of any model parameter will eventually result in divergence of orbits starting nearby each other. However, from earlier 90's, the synchronization of chaotic systems has been achieved (see for instance, [19]).

The *chaos suppression problem* is generally considered as the stabilization of the chaotic system in a periodic orbit or an equilibrium point. The chaos suppression is not an easy task if we think that a given chaotic system must be controlled despite modeling errors, parametric variations, perturbing external forces, noisy measurements and nonmodeled actuator dynamics. Moreover, only a few number of states of the chaotic systems are available from measurements (for instance, position). But, if the system to be controller is not modeled (or its model is complex). The chaos suppression could not be an easy task.

Nevertheless, many feedback strategies have been reported in the literature. Lyapunov methods [20], adaptive strategies [21], [22], time-delay coordinated schemes [23], chaos control via reconstruction of invariant manifolds [7] and robust asymptotic linearization [24]. Lyapunov methods are based on rigorous mathematical proofs and have solid fundamentals on differential geometry. Our proposal consists in a simple feedback scheme, which has robustness properties.

The class of nonlinear systems to be studied are described by the following equation

$$\ddot{x} = \zeta(x,t;\theta) + u \tag{5.15}$$

where $x = (x,\dot{x})^{\mathrm{T}}$ is a vector whose components are the position, x, and the velocity, x, of the oscillator. The terms $\zeta(x,t;\theta)$ is a nonlinear function which could be deterministic or stochastic. The input $u = u(x,t)$ is an external force used to control the oscillator (Note that this input was represented by the control torque, τ_2) and $\theta \in \mathbb{R}^p$ is a set of system parameters.

First, we have restricted ourselves to the class of second-order systems (5.15), the control methodology presented in next paragraphs can be easily extended to the class of higher-dimensional systems given as $x^{(n)} + \Sigma_{k=1}^{n-1} c_k x^{(k)} + \zeta(x,t;\theta) = \vartheta(x,t;\theta) + u$, where $x^{(i)} = d^i x/dt^i$, $i = 1,2,\ldots,n$. The terms $\vartheta(x,t;\theta)$ represents a time function, which is a bounded-away from zero and the coefficients c_k are constant. Let $x_1 = x$ and $x_2 = \dot{x}$. Then, the system (5.15) can be rewritten as $\dot{x}_1 = x_2$; $\dot{x}_2 = \zeta(x,t;\theta)+u$. Note that the above system has the form of the second-order driven oscillators, whose control were discussed previously. This implies that, the Duffing, van der Pol and Coulomb oscillators, the driven pendulum and magnetic levitation belong to the class of oscillators given by (5.15) [23], [25]. In seek of clarity, the self-organization question can be stated as follows: *Given a desired trajectory* $r(t) = (r, \dot{r}(t))^{\mathrm{T}}$, *is there a feedback law,* $u(x,t)$, *such that* $\dot{x}(t)$ *and* $x(t)$ *track asymptotically* $r(t)$ *and* $\dot{r}(t)$, *respectively, despite modeling errors, noisy measurements, parameter variations and time-lacks in the actuators?*. Note that, in such case the synergetics problem can be interpreted as control one, where the control objective is to find a feedback function, which stabilizes the oscillator around an desired reference. It should be noted that in some cases (for example, synchronization of dynamical systems), the desired trajectory $r(t)$ can be implicitly given in terms of the dynamics of a reference model. In this way, if case, the synchronization of the system (5.15) with the reference is yielded by a dynamical system of the form

$$\begin{aligned} \dot{z}_1 &= z_2 \\ \dot{z}_2 &= \xi(z,t;\phi) \\ y &= z_1 \end{aligned} \tag{5.16}$$

where $r = y$ and y is the output of the reference model. $\boldsymbol{\phi}$ is a set of reference model parameters and $\xi(z,t;\boldsymbol{\phi})$ is a nonlinear function. In particular if $\xi(z,t;\boldsymbol{\phi}) = \zeta(x,t;\theta)$, the idea of synchronizing the dynamics of the system (5.15) to the dynamics of the model is a matching problem [26]. But, in general, the synchronization problem is more interesting if $\xi(z,t;\boldsymbol{\phi}) \neq \zeta(x,t;\theta)$.

As above sections, first an ideal feedback is designed toward the final controller. Thus, let us assume that: (a) The nonlinear function $\zeta(x,t;\theta)$ is known and (b) The all states (x_1, x_2) are available for feedback. These assumptions are not realistic; however, we use them as the intermediate assumptions toward the robust controller which will be designed in the next section. The control objective s to design a control law $u(x,t)$ such that $\lim_{t\to\infty} e(t) \to 0$, where the control error $e(t)$ is defined as $e(t) = r(t) - x(t)$ and $e(t) \in \mathbb{R}^2$. One way to design the controller was presented in above sections and is constructed through a feedback control law that induces a stable linear dynamical behavior in the tracking error

$$\ddot{e} + K^T e = 0 \tag{5.17}$$

where $\ddot{e} = \ddot{r} - \ddot{x}$, $K = [K_1, K_2]$, K_1 and K_2 are design coefficients whose values are chosen in such a way that the polynomial $P_2(\lambda)$ has its roots on the open left-half complex plane (for more details see above section or [23]). Then, the feedback law given by IFC asymptotically steers the system trajectories to a behavior with $x(t) = r(t)$ and $\dot{x}(t) = \dot{r}(t)$. In fact, the IFC counteracts the nonlinear term $\zeta(x,t;\theta)$ and induces a stable linear behavior given by $V(y)$.

5.3.1 A Robust Recursive Feedback Law

Since the term $\zeta(x,t;\theta)$ is not known (Assumption *S.2.1*) and the velocity is not available for feedback (Assumption *S.2.2*), the IFC is not physically realizable. An alternative is to use estimates of $\zeta(x,t;\theta)$ and x in such a way that the main characteristics of the IFC can be retained [22], [23]. Following ideas by Ostojic [27], one can define the *desired tracking-error* dynamics as follows

$$\sigma = \ddot{e} + K^T e \tag{51.8}$$

The above equation can be rewritten in the following form $\sigma(u) = \ddot{x} - \zeta(x,t;\theta) - u + K^T e$. Since σ explicitly depends on u, we can find the control input that imposes the desired dynamics on the tracking error by solving the equation $\sigma(u) = 0$ for all $u(t)$. This means that, it is possible to reduce the control problem to that of finding the solution $u^*(t)$ for the following unsteady state equation

$$\sigma(u^*(t),t) = 0 \tag{5.19}$$

In general, the analytical solution for equation (5.19) is given by

$$u^* = \Xi(\ddot{r}, x(t), e(t), t) \tag{5.20}$$

where $(\ddot{r}, x(t), e(t), t)$ is a nonlinear map. Now, a methodology based on the *Fixed Point Theorem* (FPT) can be used to solve (5.19) (for more details, see [28]). Suppose that the desired error-tracking dynamics can be posed as follows

$$u(t) = F(u(t),t) \tag{5.21}$$

Thus, using the results reported in [28], it is possible to obtain a recursive formula for the control command

$$u(t_{k+1}) = F(u(t_k),t_k) \tag{5.22}$$

where $u(t_{k+1}) \in [u_{min}, u_{max}]$ are the bounds of the feedback and $t_{k+1} = t_k + \Delta$. The feedback (5.22) is a *Robust Recursive Feedback Law* for the system (5.15), where $u(t_0)$ is given arbitrarily and $k = 1,2, \ldots, k_\infty$. This implies that the system (5.15) tracks a derided reference signal. It must be pointed out that the control command remains bounded all the time and its value is calculated using delayed input information. We have the following sufficient conditions to assure boundedness of the estimation error $\varepsilon = u^*(t) - u(t_k)$ [27]:

(C.5.1) Both $F(u,t)$ and $u^*(t)$ are continuously differentiable functions of their arguments.

(C.5.2) There exists a constant M such that $|\partial F(u,t)/\partial u| = M < 1$ holds in a region containing

$$u(t_0), u(t) \text{ and } u^*(t) \text{ for all } t \in [t_0, t_0+\Delta k_\infty].$$

Both conditions (C.5.1) and (C.5.2) depend on the way that $F(u,t)$ is chosen. Thus, through *Successive Substitutions Method* [29], the desired tracking error dynamics can be transformed from (5.18) into the following form

$$F(u(t_k),t_k) = u(t_k) + \kappa\sigma(u(t_k),t_k) \tag{5.23}$$

where κ is an arbitrary constant parameter. Then, the *Robust Recursive Feedback Control* law (5.23) is given by

$$u(t_{k+1}) = u(t_k) + \kappa\sigma(u(t_k),t_k) \tag{5.24}$$

It is easy to show that the condition (C.5.1) is satisfied. On the other hand, the condition (C.5.2) is satisfied if

$$\left|\frac{\partial F(u,t)}{\partial u}\right| = \left|\frac{\partial}{\partial u}\left[u + \kappa\left[\ddot{x} - \zeta(x,t;\theta) - u + K^T e\right]\right]\right| = |1-\kappa| < 1 \tag{5.25}$$

Then, (C.5.2) is satisfied if $-2< \kappa< 0$. We conclude that, if $F(u(t_k),t_k)$ is given as in (5.23), the feedback (5.24) is globally stable for all $\kappa \in (-2,0)$. That implies that $\kappa \in (-2,0)$ is an adjustable parameter which can be chosen in such a way the *Robust Recursive Feedback Controller* (5.24) converges to the solution (5.20) (in this sense, the constant κ is a measure of the estimation rate).

In order to implement the controller (5.24), it is necessary to evaluate the desired tracking-error dynamics $\sigma(u,t)$. To this end, let us define e^(t) as the estimated value of e(t). Then, we propose the following discrete estimator [30]

$$\begin{aligned} \dot{\hat{e}}(t_k) &\approx \frac{(r(t_k)-x_1(t_k))-(r(t_{k-1})-x_1(t_{k-1}))}{\Delta} \\ \ddot{\hat{e}}(t_k) &\approx \frac{(r(t_k)-x_1(t_k))-2(r(t_{k-1})-x_1(t_{k-1}))+(r(t_{k-2})-x_1(t_{k-2}))}{\Delta} \end{aligned} \tag{5.26}$$

where Δ is the sampling period. With these estimates, the recursive feedback control law is given in the following form

$$\begin{gathered} \hat{u}(t_{k+1}) = u(t_t) + \kappa\left(\ddot{\hat{e}}(t_k) + K_1\dot{\hat{e}}(t_k) + K_2 e\right) \\ -2 < \kappa < 0 \end{gathered} \tag{5.27}$$

Once again, note that in the limit $\Delta \to 0$, the estimated value and its time derivatives converge to the current values. Additionally, $\hat{\boldsymbol{u}} \to u^{\mathrm{I}}(t)$. Now, let us suppose that the values for κ, K_1 and K_2 are chosen as -1,2 and 1, respectively. For this set of values, the controller proposed in [23] (which is exactly equal to the finite differences approach). This implies that the linear behavior induced by the IFC (5.6) can be recovered as $\Delta \to 0$. Moreover, the lower the sampling interval Δ is given by the lower the difference between $u^{\mathrm{I}}(t)$ minus $\hat{\boldsymbol{u}}$. In other words, the proposed controller in the above section is a particular case of the control law (5.26). Besides, note that the proposed controller only requires measurements of the position at time t_k, t_{k-1} and t_{k-2} (third-order compensator).

Now, we will show that the control strategy (5.25), (5.26) can be used to address of problems of suppression and synchronization of chaos. In fact, we will illustrate via numerical simulations that the previously developed control strategy is able to synchronize chaotic oscillators with the only knowledge of the measured state x.

Example 5.3. *Synchronization against master/slave mismatches.* It is possible to synchronize two strictly different chaotic oscillators (see [23]). Figure 5.10 shows the mean absolute error between the slave and master oscillators as a function of the value of the parameter, κ. Note that for values of the parameter κ closer to zero, the mean absolutes error is larger than the corresponding for $\kappa = -1.0$. For values of κ smaller than -1.0, the mean absolute error displays an increasing behavior. This is due to the fact that the synchronization takes a longer time. This result is in accordance to Nakamura's results[29], which point out that the convergence rate is almost optimal when $\kappa \approx (\partial F(u,t)/\partial u)^{-1} = -1$.

A general approach to control a class of second-order system systems was developed. Given a reference signal and output measurements, the controller comprises a discrete-time states and uncertainty estimator and a linearizing-like

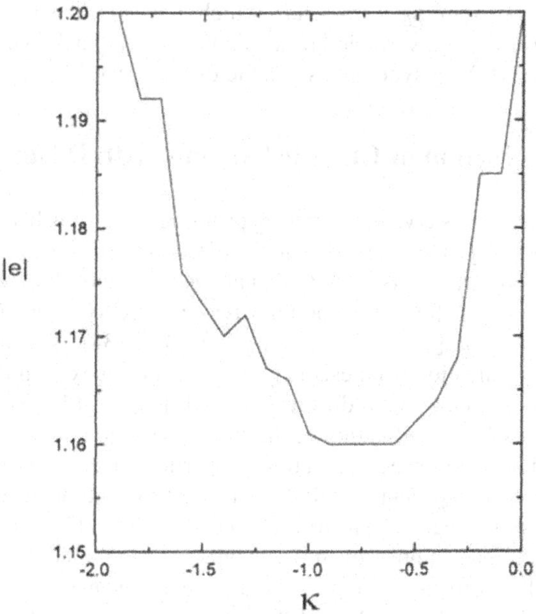

Fig. 5.10. Mean absolute error for the Duffing synchronization

feedback. By exploiting the fixed point properties of contraction maps, the stability properties of the resulting closed-loop system and control tuning guidelines were established . The control strategy was used to address the important problems of chaos suppression and synchronization. Numerical simulations were provided to illustrate the performance of the controller. The order conjecture was proved for the discrete-time controller. This is, a chaotic system, of order n, can be controlled by means of a feedback whose order is n+1. Such controller can leads the trajectories of the chaotic system to a desired reference. The reference signal can be provided by an arbitrary choice (chaos suppression) or by the output of an arbitrary dynamical system (chaos synchronization).

The feedback studied has very simple structure and allows chaos synchronization even if the master and slave have strictly different model. In some sense, the feedback scheme attains the organization of both master and slave oscillators vi the coupling parameter, u (which is so-called control command). Such coupling parameter is the results of the feedback interconnection between systems. This results leads us to think that self-organization rules have feedback structure. In fact, discrete-time feedback induces a geometrical transformation into the system (see Section 5.1). Therefore, it seems the synergetic mechanism.

Nevertheless, there are many questions. For example: (I) Can simple feedback laws induce synchronization even if master and have not the same order?. For instance, the heart could be a different order system than lung. However, they are, in some way, self-organized. (II) Can simple feedback laws induce self-organization even in presence of unobservable states?. Let us suppose that heart and lung are same order system. It is obvious that they are self-organized. The point is: Both heart and lung are interconnected via coupling parameter, which have feedback structure. One can expect that only few state variable are available for feedback. Then, a feedback law could allow the heart-lung synergetics in spite of unobservable states.

5.4 Synchronization of Chaotic Systems with Different Order

Chaos synchronization is a very interesting problem which has been widely studied in recent years. Actually, there are two main directions in the research of the chaos synchronization: (i) synthesis and (ii) analysis. The problem of synchronization synthesis is to design a force for coupling two chaotic systems. The coupling force can be designed linear [31], [32] or nonlinear [33], [34]. The actual challenge in problem of synchronization synthesis is to achieve and to explain the synchronization between chaotic systems with different model [35], [36], [37]. Synchronization analysis consists in (a) classification of the synchronization phenomena [38], [39] and (b) comprehension of the synchronization properties (such as robustness [40] and/or bifurcation [41). An important challenge in synchronization analysis is to develop genuine indicators of chaotic synchronization. Such a problem arises from classical indicators failure. For example, even if Lyapunov table indicates that chaos has been controlled, small disturbances can provoke deficiency in such indicators [42].

This work addresses the problem of the synchronization synthesis. Such a problem can be classified in the following research areas: (i) the study of potential applications for chaos synchronization and (ii) development of synchronization strategies. Concerning the first research area, chaos synchronization has application in several fields as biological systems, where the research is focused in neurons lattices [43], [44], and transmissions of secure message [45], [46]. Regarding development of synchronization schemes, several strategies can be found in literature (for example linear [32], nonlinear [34] or adaptive [47]). In particular, a synchronization strategy is developed in this paper from nonlinear feedback. Synchronization synthesis is a problem of first generation while applications of synchronization is a problem of second generation. However, there is no full knowledge about the synchronization synthesis. Thus, for instance, theoretical and experimental results show synchronous behavior in nature. For instance, in [41] and [43], authors have shown very interesting results about the chaotic synchronization phenomenon in neurons. Nevertheless, *on the microscopical scale the mutual interactions are not yet clearly understood.* Indeed, only phenomenon models have been used [42]. In principle, it is possible that synaptic communication can be yielded between neurons with different dynamic model. For example, synchronous activity has been observed in thalamic and hippocampal neurons networks [48]. An alternative for understanding synchronous behavior in nature is to develop synchronization strategies. In articular, a synchronization strategy is developed in this paper.

There is one interesting question in this direction: is the synchronous behavior in nature yielded from a feedback of a fed-forward coupling force? There is no definitive answer for above question. As a matter of fact, synchronization is the results of the coupling between dynamical systems. The coupling can be performed via two basic interconnections: fed-forward and feedback. Fed-forward interconnection consists of the input of a dynamical signal without return. On contrary, feedback implies that a potion of the system is returned. Both interconnections have been used for chaos control; see [49] and [50] for fed-forward and [32-36,38] for feedback. Indeed, a combination of both interconnections can be performed for synchronization of chaotic oscillators [51]. Since both interconnections achieve chaos synchronization, one is unable to confirm that synchronous behavior in nature is only a consequence of feedback coupling. However, synthesis of chaos synchronization via feedback allows us to expect promissory results. In this sense, some results show that synchronization between chaotic systems whose model is strictly different has been reported [35-37]. Nevertheless, until our knowledge, there is no previous results about the synchronization between chaotic systems whose order is different. Such a problem is reasonable if, for instance, we think that order of the thalamic neurons can be different than hippocampal neurons [48]. One more example is the synchronization between heart and lung. One can observe that both circulatory and respiratory systems behave in synchronous way. However, one can expect that model of the circulatory system is strictly different than respiratory system, which can involve different order. In addition, the synchronization of strictly-different chaotic systems is interesting by itself.

In this section we deal with the synthesis of the chaos synchronization between oscillator with different model. We present a nonlinear approach for synchronizing chaotic system whose order is not equal. In particular, we present the matching of the Duffing equation attractor with a projection onto one canonical plane of the Chua circuit (double-scroll oscillator). In this sense, results show that *reduced order synchronization* can be achieved by nonlinear systems. This means that two chaotic systems can be synchronized in spite of order of the response system is less than order of the drive system. Of course, since order of response oscillator is smaller than master system, the synchronization is only attained in reduced order. Results are focused on geometrical features of synchronization phenomenon. Chua-Duffing synchronization is allowed by a nonlinear feedback, which yields a smooth and bounded coupling force (i.e., control command). The nonlinear feedback is designed from a simple algorithm based on time derivative of system output along the master/slave vector fields.

5.4.1 Problem Statement

Chua system is an electronic circuit with one nonlinear resistive element. The circuit equations can be written as a third order system which is given by the following dimensionless form [52]

$$\begin{aligned} \dot{x}_1 &= \gamma_1(x_1 - x_2 - f(x_1)) \\ \dot{x}_2 &= x_1 - x_2 + x_3 \\ \dot{x}_3 &= -\gamma_2 x_2 \end{aligned} \tag{5.27}$$

where $f(x_1) = \gamma_3 x_1 + 0.5(\gamma_4 - \gamma_3)[\,|x_1 + 1| - |x_1 - 1|\,]$, γ_i's, i = 1,2,3,4, are positive constant. Let us assume that system (5.27) represents the drive system. Figure 5.11.a shows the phase portrait of system (5.27) and its projection on the canonical planes. The parameters were chosen as follows: $\gamma_1 = 10.0$, $\gamma_2 = -14.87$, $\gamma_3 = -0.68$ and $\gamma_4 = -1.27$. Initial condition were arbitrarily located at the point $x(0) = (0.1,-0.5,0.2)$.

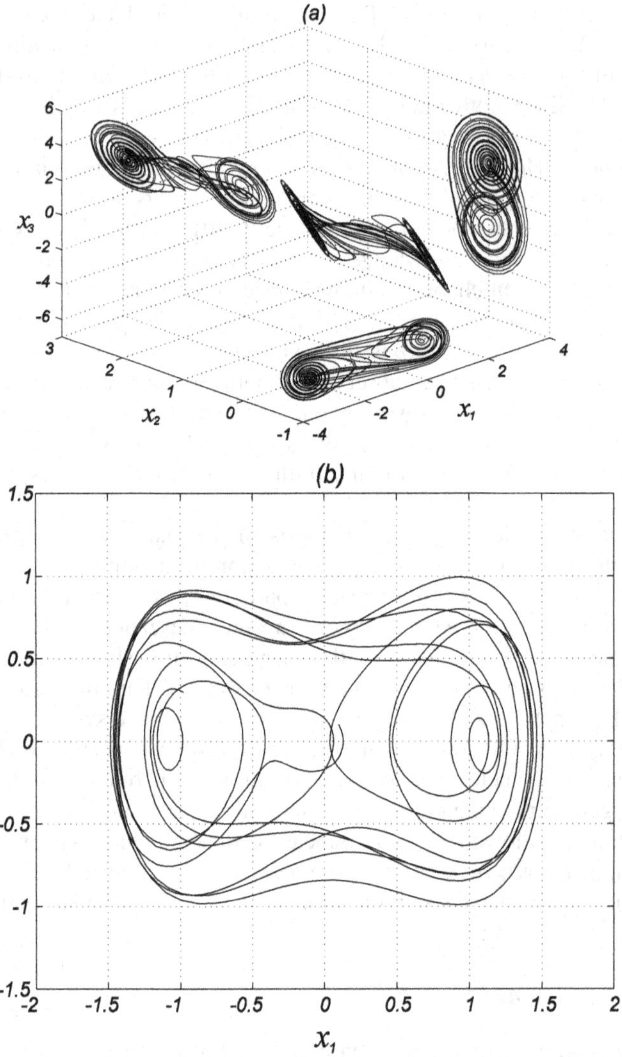

Fig. 5.11. Phase portrait of the Chua circuit and Duffing equation without coupling. (a) attractor of the Chua system and its projections on canonical planes. (b) attractor yielded by Duffing equation.

Now, let us consider the Duffing equation, which is given by

$$\begin{aligned} \dot{x}'_1 &= x'_2 \\ \dot{x}'_2 &= x'_1 - x'^3_1 - \delta x'_2 + \tau_e(t) + u \end{aligned} \tag{5.28}$$

where δ is a positive parameter which represents damping coefficient, $\tau_e(t) = \alpha\cos(\omega t)$ denotes driving force and u is the coupling force (controller). Figure 5.11.b shows the phase portrait of Duffing equation for $u = 0$ for all $t \geq 0$ (uncontrolled evolution).

Parameters were chosen as $\delta = 0.15$, $\alpha = 0.3$ and $\omega = 1.0$. Initial conditions were arbitrarily located at origin.

Obviously, there is difference between attractors of systems (5.27) and (5.28) (see phase portrait in Figure. 5.11.b and projections in Figure 5.11.a), i.e., both systems (5.27) and (5.28) are not synchronized neither phase nor frequency. Moreover, there is no synchronization in any sense (see [38], [39] for details concerning definition of synchronization kinds). The classical synchronization problem is somewhat distinct than problem of synchronizing different chaotic systems. In classical synchronization problem, drive and response system has similar geometrical and topological properties [53]. Thus master/slave interconnection can be sufficient to attain synchronization [54]. Latter one is understood as adjustment of master/slave dynamics due to coupling via output of both systems. Synchronization of different chaotic systems is a hard task if we think that: (i) initial conditions of master and slave systems are different and unknown, (ii) topological and geometrical properties of different chaotic systems are quite distinct and (iii) unrelated chaotic systems have strictly different time evolution.

Systems (5.27) and (5.28) have been widely used to study chaos synchronization when drive and response system have same order. However, we are interested in synchronization of Duffing attractor with a projection in a canonical plane, *i.e.*, to lead the attractor of equation (5.28) to the geometrical properties of a projection of system (5.27). To study such a problem, we choose x_3 as measured state from master system (Equation (5.27)) whereas $x'1$ is the measured observable of the slave system (Equation (5.28)). Thus the chaos control objective is *to design a feedback u such that the discrepancy error* $e = x_1' - x_3$ *tends to zero as* $t \to \infty$. Note that above goal implies that if $e \to 0$ as $t \to \infty$, then $x_1' \to x_3$ for all $t \geq t_0 \geq 0$ and any initial discrepancy $e(0) = x_1'(0) - x_3(0)$. Therefore, at least, the partial synchronization [38], [39] of both master and slave systems can be attained via feedback coupling. However, the control objective is to achieve the reduced order synchronization, *i.e.*, all states of the response system should be synchronized, in some sense, with any states of the drive system.

5.4.2 The Proposed Feedback and Design Details

Synchronization of chaotic systems via feedback can be addressed from the stabilization of the synchronization system error around origin (see for instance, [36]) or as the tracking of an target trajectory. First case consists in construction of the dynamical systems of the synchronization error in such way that the feedback scheme leads its trajectories to origin. The interesting problem is to attain the synchronization objective (*i.e.*, stabilization around the origin) in spite of the synchronization error system is uncertain. Latter one comprises the leading of the slave trajectories to the master ones (e.g., to lead $x'1(t)$ to trajectory $x_3(t)$). That is, in second case the goal is to direct the slave system to the desired trajectory (which is provided by the master system). This problem has been recently addressed (see for instance [53]). However, when synchronization is not solved at all when it is addressed as a tracking problem due to tracking cannot be achieved by simple feedback [54]. On contrary, stabilization of the synchronization error around origin is promissory [32-40]. As we shall see below, in this paper the synchronization of systems with different order is addressed as the stabilization of the synchronization error at origin.

Let us assume the following

S.4.1) Only x_3 is available for feedback from system (5.27).
S.4.2) x_1' and x_2' are available for feedback from slave system.
S.4.3) Vector fields of master and slave systems are smooth.

Now, let us consider the difference $e(t) = x_3(t) - x_1'(t)$. From first time derivative of the difference error one has that $\dot{e} = \dot{x}_3 - \dot{x}_1' \equiv - \gamma_2 x_2 - x_2'$ and second time derivative of difference error one has that $\ddot{e} = - \gamma_2 \dot{x}_2 - \dot{x}_2' \equiv - \gamma_2(x_1 - x_2 + x_3) - x_1' + x_1'^3 + \delta x_2' - \alpha\cos(\omega t) - u$. Now, from simple algebraic manipulations, the following equation is obtained: $\ddot{e} - \dot{e} + \gamma_2 e = (1 + 2\delta)x_2' + (1 - 2\gamma_2)x_1' - \alpha\cos(\omega t) - u$ or equivalently

$$\begin{aligned} \dot{e}_1 &= e_2 \\ \dot{e}_2 &= e_1 - \gamma_2 e_2 - \tau'_e(x', t; p) - u \end{aligned} \tag{5.29}$$

where $\tau_e'(x', t; p) = (1 + 2\delta)x_2' + (1 - 2\gamma_2)x_1' - \alpha\cos(\omega t)$ can be interpreted as a disturbance force acting onto the linear system (5.29), $p \in \mathbb{R}^4$ denotes a parameter set and $x' \in \mathbb{R}^2$ is a disturbance vector. Thus, the synchronization problem becomes to find the control command u such that system (5.29) is asymptotically stable at origin for any initial condition $e(0) \in \mathbb{R}^2$. Note that if control command u is able to stabilize system (5.29) at origin, then the synchronization error and its dynamics can be leaded to zero (*i.e.*, u induces a steady state $e(t) = 0$ for all $t \geq t_0 \geq 0$, where t_0 is the time when control is activated). That is, dynamic evolution of slave system can be manipulated toward the master behavior.

5.4.3 Reduced-Order Synchronization under Partial Knowledge

We propose the following feedback

$$u = (1 + 2\delta)x'_2 + (1 - 2\gamma_2)x'_2 - \alpha\cos(\omega t) + k_1 e_1 + k_2 e_2 \tag{5.30}$$

where k_1, k_2 are constant parameters of the control which are computed from the following procedure. Equation (5.29) under controller (5.30) action results in the following closed-loop system: e = Ae where the matrix A is given by

$$A = \begin{bmatrix} 0 & 1 \\ 1 - k_1 & -(\gamma_2 + k_2) \end{bmatrix} \tag{5.31}$$

Then k_1 and k_2 are chosen such that matrix has all its eigenvalues at the open left-hand complex plane (i.e., all roots of $\lambda^2 + (\gamma_2 + k_2)\lambda + 1 - k_1 = 0$ have negative real part), which is satisfied for any $k_2 > \gamma_2$ and $0 < k_1 < 1.0$.

Note that controller (5.30) yields linear behavior into system (5.29). That is, integration of system (5.29) under controller (5.30) yields $e(t) = e(0)\exp(At)$, where $e(0) \in \mathbb{R}^2$ is the initial condition of the synchronization error and A is a stable matrix for a given k_1 and k_2. Feedback based on equation (5.30) has two advantages: (i) it does not require full knowledge of the system (5.27). Indeed, feedback (5.30) allows coupling between system (5.27) and (5.28). Such a design is reasonable. For example,

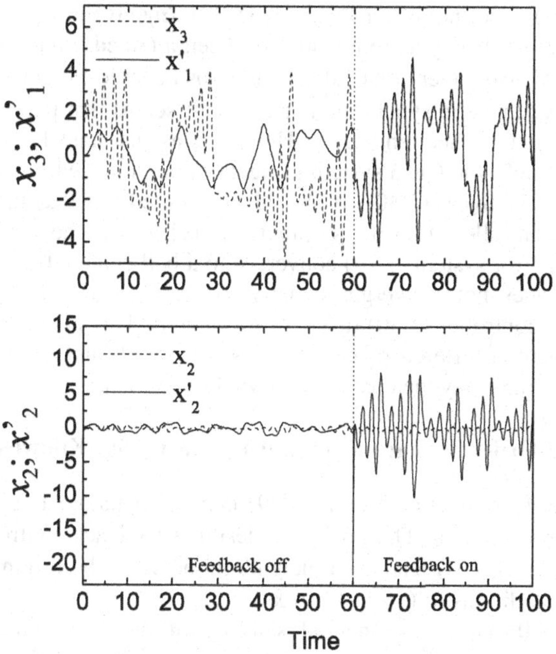

Fig. 5.12. The difference $e = x_3 - x_1'$ holds close to zero under feedback actions. Time evolution of $x_2(t)$ has different amplitude than $x_2'(t)$.

in neural systems, different neurons in one subsystem are always driven by output from neurons in higher level. Such differences can be interpreted as the time series of strictly different dynamical systems. Thus differences between signals in higher level neurons can be interpreted as the time series of strictly different dynamical systems. Hence, synchronization of dynamical systems from such time series plays an essential role in nonlinear processes. Thus, if master and slave systems have distinct properties, it is expected that direct interaction between them cannot necessarily yield synchronization. It is our belief that synchronization in nature is given by several kinds of interactions by feedback coupling. (ii) it compensates the nonlinear terms and induces a linear behavior. Indeed, this is the main desirable feature of nonlinear control.

Nevertheless, the controller (5.30) has the following drawbacks: (i) the stability of the system (5.29) under controller (5.30) is based on the choice of the eigenvalues of the matrix (5.31). This procedure is known as pole placement in control theory. Pole placement can result in poor control actions and (ii) controller (5.30) requires the values of the slave parameters.

The crux is obvious, what happen if the parameters δ and γ_2 are time-varying or are not exactly known? Both drawbacks will be discussed in next subsection. However, it is pertinent to illustrate that feedback (5.30) yields the reduced-order synchronization. To this end, and in seek of clarity, some numerical simulations were performed. Without lost of generality, one can consider that parameters and initial conditions have same value than Figure 5.11 (see above). Figure 5.12 shows time evolution of

(x_2,x_3) and (x_1',x_2') under nonlinear feedback (5.30). Eigenvalues of matrix (5.31) were located at -30. Note that synchronization has been attained. Figure 5.12.a shows that states x_3 and x_1' evolve under practical synchronous behavior. Figure 5.12.b shows that x_2 and x_2' do not behave in synchronous manner. It seems that partial synchronization has been attained (see Section 3.2.1). However, as we state below, trajectories of duffing system tracks the Chua system projection, i.e., reduced-order synchronization is performed by feedback (5.30). It should be pointed out that the synchronization achieved by controller (5.30) is practical [9]; i.e., the trajectories of the synchronization error system (5.29) converges to a ball centered at origin. This means that error $e(t)$ does not converges exactly to zero. As a consequence, the time-evolution of the controller (5.30) does not go to zero. This is due to feedback (5.30) absorbs the structural differences between systems (5.27) and (5.28). In other words, feedback (5.30) "must pay the cost" of the synchronization.

5.4.4 Reduced-Order Synchronization with Least Prior Knowledge

As was discussed above, the system (5.29) can be, if parameters are unknown, an uncertain nonlinear system. The goal is to design a feedback controller such that the synchronization error $e_1(t)$ tends to a neighborhood Ω of the origin for all $t \geq t_0 \geq 0$ and any initial conditions $(e_1(0),e_2(0))$ in $\mathbb{R}^2$.

Feedback (5.30) yields exponential stability of the synchronization error, $e(t)$. However, it is quite complex and requires a priori information about slave oscillator. Then, a modification is desirable in such manner main features of feedback (5.30) be held.

Let us define $\eta = \gamma_2 e_2 + \tau'_e(x',t;p)$ as a new variable, which is smooth. Of course, if $\tau'e(x',t;p) = (1 + 2\delta)x_2' + (1 - 2\gamma_2)x_1' - \alpha\cos(\omega t)$ is uncertain, the state η is not available for feedback. We propose the following procedure for designing the feedback controller with least prior knowledge. The uncertain term η can be computed from equation (5.29) in the following way: $\eta = \dot{e}_2 - e_1 + \boldsymbol{u}$. Now, by approaching the time derivative by means of finite differences at time $t_k \in [t_k,t_{k+1})$, one has that $\eta(t_k) \approx \hat{\eta}(t_k) = -e_1(t_k) + u(t_{k-1}) + (e_1(t_k) - e_1(t_{k-1}))/\Delta t$, where Δt denotes the sampling rate. Such an approach provides an estimated value of the uncertain term η at time t_k from the measurements of the error $e(t_k)$ and the last control input $u(t_{k-1})$. In this way, the controller (5.30) becomes

$$u(t_k) = -\hat{\eta}(t_k) + k_1 e_1(t_k) + \frac{k_2}{\Delta t}(e_1(t_k) - e_1(t_{k-1})) \tag{5.32}$$

where k_1 and k_2 are chosen such that matrix (5.31) has all its eigenvalue located at open left-hand complex plane. Note that proposed controller comprises two parts: (i) the feedback (5.32) and (ii) the uncertainties estimator given by the finite differences approach. In addition, such a controller does not require prior knowledge neither the parameters values nor model of the master system. In principle, as $\Delta t \to 0$ as the estimated value $\hat{\eta}(t_k) \to \eta(t)$ for all $t \in [t_k,t_k + \Delta t) \geq t_0 \geq 0$, where t_0 denotes the time where controller is activated. Hence, as $\Delta t \to 0$, the controller (5.32) will behave as the nonlinear controller (5.30). This means that as $\Delta t \to 0$, nonlinear terms of the synchronization system (5.29) can be counteracted by the controller (5.32). It should be pointed out that if $\Delta t = 0$, then the controller (5.32) is not physically realizable. This makes sense because $\Delta t = 0$ means "no sampling rate". However, such a condition implies that noise sensitivity can be displayed by controller (5.32). Figure 5.13.a shows the dynamics of the synchronization error under controller (5.32) for $\Delta t = 2.5$ Hz. The time evolution of the respective control command is shown in Figure 5.13.b. Here, same values of the control parameters $k_1 = 25.0$ and $k_2 = 10.0$ were chosen for controller (5.30) and (5.32). This implies that all eigenvalues of the matrix (5.31) are located at -5.0. Note that, in spite of eq. (5.32) requires least prior knowledge about the synchronization system, the proposed strategy is able to achieve reduced-order synchronization for relatively small sampling rates. Indeed, as smaller sampling rate as better synchronization. In addition, Figures 5.13.a and 5.13.b shows that synchronization under feedback (5.30) and (5.32) are equivalent.

5.4.5 Discussion of the Results

In this section, we briefly discuss the obtained results. The Figure 5.14 shows the projection of Chua system (eq. (5.27)) on (x_2,x_3)-plane and phase portrait of the Duffing oscillator under feedback actions (eq. (5.28)).

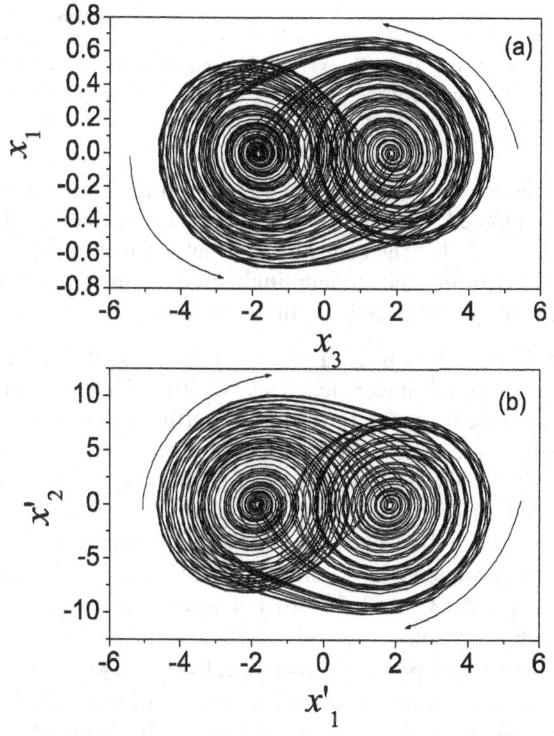

Fig. 5.13. Reduced-order synchronization under feedback (5.32). (a) Duffing equation yields an attractor which is a mirror reflection of (b) the (x_2,x_3)-plane projection of the Chua system.

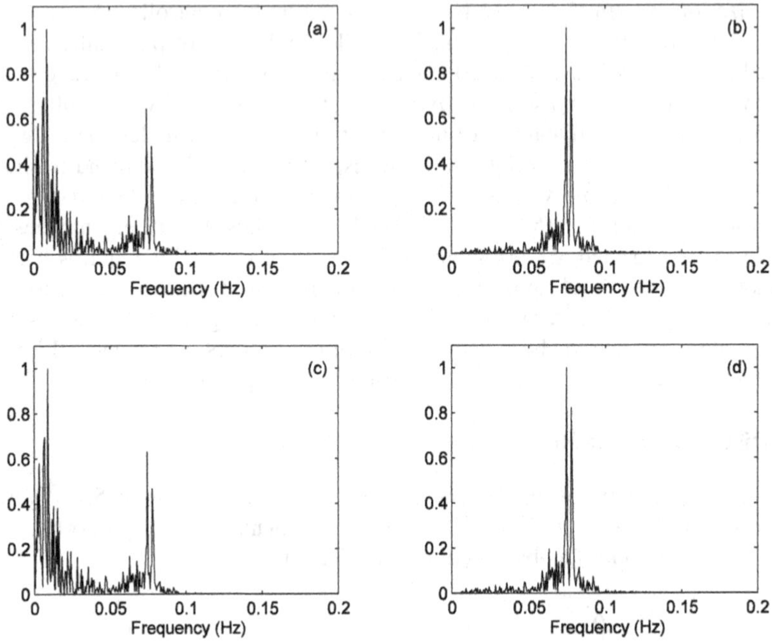

Fig. 5.14. Power spectrum of (a) x_3, (b) x_2, (c) x_1' (under control actions) and (d) x_2' (under control actions). Note that spectrum in Duffing equations (under control actions) are similar than Chua system (x_2,x_3)-plane projection.

Here, eigenvalues of the matrix (5.31) were arbitrarily located at -5 and control command was computed by the feedback (5.32) (including the uncertainties estimator) with $\Delta t^{-1} = 2.5$ Hz. The controller was activated at time $t_0 \geq 0$. Two features should be noted: (a) *chirality* and (b) amplitude oscillations. The following comments are in regarding to both chirality and amplitude features:

(a) Note that Figure 5.14.b is a mirror reflection of Figure 5.14.a. (Same phenomenon was observed under feedback (5.30)). That is, while (x_2,x_3)-plane projection of Chua system rotates toward left, Duffing equation under feedback (6) rotates toward right. Note that Duffing attractor under actions of feedback (5.32) is not *superimposable* on the mirror image of Chua (x_2,x_3)-plane projection. Such property is so-called *chirality* (see chapter 5 in [55] for introductory notion). The chirality notion has been burrowed from study of organic molecules. Thus, for example, notice that symmetry axis corresponds to state x_1'. That is, in x_1' direction both attractor (Figures 5.13.a and 5.13.b) are equal (see Figure 5.12.a) where the reflection is in the direction of x_2' is chiral.

Now, one can expect that power spectrum peaks of x_2 and x_3 are equal than x_1' and x_2', respectively. Hence reduced-order synchronization of systems (5.27) and (5.29) is attained in phase and in a practical sense [38], [39]. Although these results are novelty, they are a consequence of how the synchronization is addressed. Thus, chirality cannot yet be claimed as a kind of synchronization; however, this feature

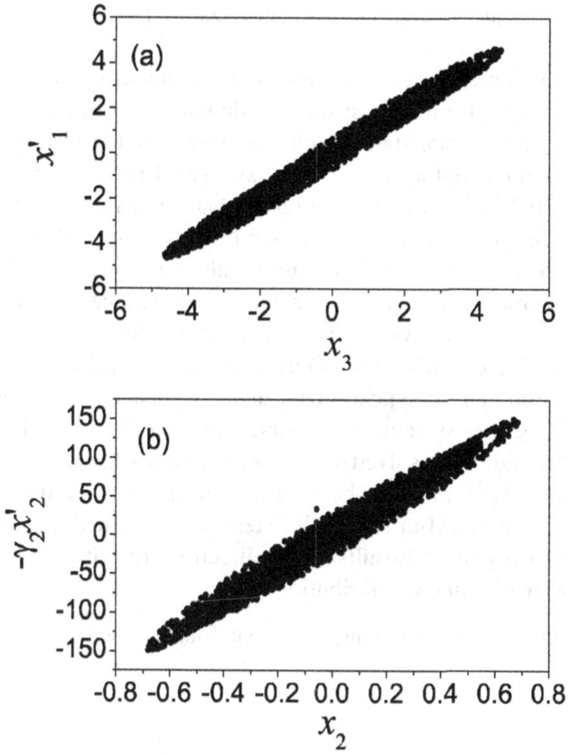

Fig. 5.15. (a) Phase locking diagram of the synchronization error $e_1(t) = x_1'$-x_3 variables. (b) Modified phase locking diagram for state variable in synchronization.

requires a detailed study. Figure 5.15 shows the PSD of x_2, x_3, x_1' and x_2'. PSD has been normalized by maximum amplitude peak. Such a picture shows that, at least, frequency and phase synchronization is attained by feedback (5.32). PSD is an important measure of synchronization. Although, PSD is not sufficient to conclude that synchronization exists; however, it is a good evidence [38].

(b) Concerning the magnitude of the attractors, we can note that, under control actions, the following steady state is obtained: $e(t) \rightarrow 0 \Rightarrow \| e(t) \| = \| x_3(t) - x_1'(t) \| \approx 0$ for al $t > t^+ \geq 0$, where t^+ is the time of control activation. Such a steady state implies that all time derivatives of e are zero for all $t \geq 0$. That is, $\dot{\boldsymbol{e}} \equiv \mathbf{0}$ for all $t \geq 0$, which implies that the two norm $\|x_2'(t)\| = \gamma_2^{-1}\|x_2(t)\|$. Therefore, at steady state, an amplification factor γ_2 affects the phase portrait of controlled Duffing equation.

However, as was stated above, synchronization is achieved. In order to add evidence, we have plotted x_3 versus x_1' and x_2 versus - $\gamma_2 x_2'$ (which can be seen as a modified phase locking diagram, see Figure 5.15). Since x_2' essentially has similar dynamic properties than x_2 (see for instance Figure 5.14). It should be pointed out that the dynamics of the synchronization error is close to zero. This means that controller

(5.32) provides practical reduced-order synchronization [39] of the drive and response systems.

Thus the reduced-order synchronization is the problem of synchronizing a slave system with projection of a master system. It should be noted that the reduced-order synchronization is not partial-state synchronization. On the one hand, partial-state synchronization is for coupling two chaotic systems whose order is equal. A main feature of the partial-state synchronization is that, at least, one state of the slave system is not synchronous in some sense (see [39], [51] and references therein). On the other, in reduced-order synchronization, all states of the slave system are synchronous, in quiral sense. The main feature of the reduced-order synchronization is that order of the slave system is less than master one. In this sense, the synchronization of Chua oscillator and Duffing equation is achieved in reduced order, *i.e.*, Duffing equation can be synchronized under feedback actions with a canonical projection of the Chua system. Of course, the problem of the reduced-order synchronization has not been solved yet. Some questions have been opened: (i) Can the reduced-order synchronization be achieved from fourth order and third order systems? (ii) Can the reduced-order synchronization be attained from fourth order and second-order driven systems?. Results in this direction are expected. Nevertheless, we belief that this paper is a timely contribution.

Exercise 5.4. Consider the following nonlinear fourth order chaotic systems as the master

$$\begin{aligned}
\dot{x}_{1m} &= x_{2m} \\
\dot{x}_{2m} &= -cx_{2m} - d^2(x_{1m} - x_{3m} - x_{1m}x_{3m}) \\
\dot{x}_{3m} &= x_{4m} \\
\dot{x}_{4m} &= -ax_{3m} - x_{3m} - bx_{3m}^3 - f(x_{1m} + \tfrac{x_{1m}^2}{2}) + E\cos\omega_3 t
\end{aligned}$$

and the Duffing system with two external forces as the slave [56], given by $\dot{x}_{1,S} = \dot{x}_{2,S}$, $\dot{x}_2 = -\lambda x_{2,S} - w_o^2 x_{2,S} - \gamma x_{1,S}^3 + K_1\cos(\omega_1 t + \theta_1) + K_2\cos(\omega_2 t + \theta_2)$. Perform the procedure developed in section 5.4.2 to obtain a discrete-time control low, in such a way that the given systems synchronize in reduce order fashion. Parameters are a = 0.03, b = 1, f = 0.1, c = 0.3, d = 0.985, E = 0.7 and $\omega_e = 1$, and consider trivial initial conditions for the master system. For the slave system initial conditions $x_o(0)$ = (0.1,0) and parameters $\lambda = 1$, $\omega_o = 10$, $\gamma = 100$, $K_1 = K_2 = 1$, $\omega_1 = 2$, $\omega_2 = 4$, and $\theta_1 = \theta_2 = 0$. □

By now, the results show an interesting phenomenon. The reduced-order synchronization yields a mirror reflection of the drive system. This can be casual; however, one more question arises: Is the mirror reflection a synchronization phenomenon or a casualty? Unfortunately, answer is beyond of the paper goal. But, if mirror reflection is a synchronization phenomenon, *chiral synchronization* can be an interesting and relevant discovery. Chiral synchronization should be characterized because the attractor of salve system is, once synchronized, a mirror reflection of the master system attractor, *i.e.*, Duffing attractor under control actions cannot be superimposed on its mirror image. The mirror reflection of the synchronization is a consequence of how the feedback control is designed.

References

[1] Armtrong-Hélouvry, B.: Control of Machines with Friction. Kluwer Academic Press, Boston (1991)

[2] Canudas de Wit, C., Noël, P., Aubin, A., Brogliato, B.: Adaptive Friction Compensation: low velocities. Int. J. Robotics Resch. 10, 189 (1991)

[3] Dupont, P.E.: Avoiding stick-slip in possition and force control through feedback. In: Proc. of the 1991 IEEE 1991 Int. Conf. Robotics Automation, Sacramento, pp. 1470–1475 (1991)

[4] Armtrong-Hélouvry, B.: Stick Slip and Control in Low-Sped Motion. IEEE Trans. Autom. Contr. 38, 1483–1496 (1993)

[5] Ot, E., Grebogi, C., Yorke, J.A.: Controling Chaos. Phys. Rev. Letts. 64, 1196 (1990)

[6] Alvarez-Ramírez, J., Garido, R., Femat, R.: Control of Systems with Friction. Phys. Rev. E. 51, 6235 (1995)

[7] Hikihara, T., Moon, F.C.: Chaotic levitated motion of a magnet supported by superconductor. Phys. Letts. A. 191, 279 (1994)

[8] Mosayebi, F., Qammar, H.K., Hartley, T.T.: Adaptive estimation and synchronization of chaotic systems. Phys. Lett. A 161, 255 (1991)

[9] Alvarez-Ramírez, J., Vargas-Vilamil, F.: State estimation for a class of nonlinear oscillators with chaotic attractor. Phys. Lett. A 197, 116 (1995)

[10] Chen, G., Dong, X.: On feedback control of chaotic continuous-time systems. IEEE Trans. Circ. and Syst. I 40, 591 (1993)

[11] Dorato, P.: Robust Control. IEEE Press, New York (1987)

[12] Aström, K.J., Wittenmark, B.: Adaptive Control. Addison-Wesley, New York (1989)

[13] Wiggins, S.: Global Bifurcations and Chaos. Springer, New York (1988)

[14] Takens, F.: Dynamical System and Turbulence. Rand, D., Young, L.S. (eds.). Springer, Berlin (1981)

[15] Esfandiari, F., Khalil, H.K.: Output feedback stabilization of fuly linearizable systems. Int. J. Control 56, 1007 (1992)

[16] Bai-Lin, H.: Chaos II. World Scientific Publishing Co., Singapore (1990)

[17] Schuster, H.G.: Deterministic Chaos. VCH Publishers, Germany (1989)

[18] Carrol, T.L., Pecora, L.M.: Synchronizing chaotic circuits. IEEE Trans. Circ. and Syst. I 38, 453 (1991)

[19] Nijmeijer, H., Berghuis, H.: On Lyapunov control of the Dufing equation. IEEE Trans. Circ. and Syst. 42, 473 (1995)

[20] Wu, C.W., Yang, T., Chua, L.O.: On adaptive synchronization and control of nonlinear dynamical systems. Int. J. of Bifurcation and Chaos 6, 445 (1996)

[21] Di Bernardo, M.: An adaptive approach to the control and synchronization of continuous-time systems. Int. J. of Bifurcation and Chaos 6, 557 (1996)

[22] Alvarez-Ramirez, J., Femat, R., Gonzalez, J.: A time delay coordinates strategy to control a class of chaotic oscillators. Phys. Lett. A 221, 41 (1996)

[23] Femat, R., Alvarez-Ramirez, J., Gonzalez, J.: A strategy to control chaos in nonlinear driven oscillators with least prior knowledge. Phys. Lett. A 224, 271 (1997)

[24] Femat, R.: A control scheme for the motion of a magnet supported by type-II superconductor. Physica D 111, 347 (1998)

[25] Aström, K.J., Witenmark, B.: Adaptive Control. Addison-Wesley, NY (1989)

[26] Ostojic, M.: Numerical approach to nonlinear control design. Trans. of the ASME 118, 332 (1996)

[27] Kincaid, D., Cheney, W.: Numerical Analysis: Mathematics and Scientific Computing. Brooks/Cole Publishing Co. (1991)

[28] Nakamura, S.: Numerical Analysis and Graphic Visualization with MATLAB. Prentice Hall Inc., NY (1995)

[29] González, J., Femat, R., Alvaez-Ramírez, J., Aguilar, R., Barron, M.: A discrete approach to the control and synchronization of a class of chaotic oscillators. IEEE Trans. on Circ. and Syst. I 46, 1139 (1999)

[30] Wu, C.W., Chua, L.O.: Synchronization in an aray of linearly coupled dynamical systems. IEEE Trans. Circ. and Syst. I 42, 430–447 (1995)

[31] Kapitaniak, T., Sekeita, M., Ogorzalek, M.: Montone synchronization of chaos. Int. J. of Bifur. and Chaos 6, 211–217 (1996)

[32] Grassi, G., Mascolo, S.: Nonlinear observer design to synchronize hyperchaotic systems via scalar signal. IEEE Trans. Circ. and Syst. I 44, 1011–1014 (1997)

[33] Nijmeijer, H., Mareels, M.Y.: An observer looks at synchronization. IEEE Trans. Circ. and Syst. I 44, 882–890 (1997)

[34] Xiaofeng, G., Lai, C.H.: On the synchronization of different chaotic oscillations Chaos. Solitons and Fractals 11, 1231–1235 (2000)

[35] Femat, R., Alvarez-Ramírez, J.: Synchronization of a clas of strictly different oscillators. Phys. Letts. A 236, 307–313 (1997)

[36] Bragard, J., Boccalletti, S.: Phys. Rev. E 62, 6346–6351 (2000)

[37] Brown, R., Kocarev, L.: A unifying framework of chaos synchronization for dynamical systems. Chaos 10, 344–349 (2000)

[38] Femat, R., Solís-Perales, G.: On the chaos synchronization phenomena. Phys. Letts. A 262, 50–60 (1999)

[39] Rulkov, N.F., Sushchik, M.M.: Robustness of synchronized chaotic oscillations. Int. J. of Bifur. and Chaos 7, 625–643 (1997)

[40] Bazhenov, M., Huerta, R., Rabinovich, M.I., Sejnowski, T.: Cooperative behavior of a chain of synapticaly coupled chaotic neurons. Physica D 116, 392–400 (1998)

[41] Aguirre, L.A., Billings, S.A.: Closed-loop suppresion of chaos in nonlinear driven oscillators. J. Nonlinear Sci. 5, 189–206 (1995)

[42] Tass, P., Haken, H.: Synchronized oscillations in the visual cortex - a synergetics model. Biol. Cybern. 74, 31–39 (1996)

[43] Huerta, R., Bazhenov, M., Rabinovich, M.I.: Cluster of synchronization and bistability in a latices of chaotic neurons. Europhys. Letts. 43, 719–724 (1998)

[44] Pyragas, K.: Transmission of signals via synchronization of chaotic time-delay systems. Int. J. of Bifur. and Chaos 8, 1839–1842 (1998)

[45] Short, K.M.: Steps toward unmasking secure communications. Int. J. of Bifur. and Chaos 4, 959–977 (1994)

[46] Mosayebi, F., Qammar, H.K., Hartley, T.T.: Adaptive estimation and synchronization of chaotic systems. Phys. Letts. A 161, 255–262 (1991)

[47] Terman, D., Koppel, N., Bose, A.: Dynamics of two mutually coupled slow inhibitory neurons. Physica D 117, 241–275 (1998)

[48] Lima, R., Pettini, M.: Suppresion of chaos by resonant parametric perturbations. Phys. Rev. A 41, 726–733 (1990)

[49] Pecora, L.M., Carrol, T.L.: Synchronization in Chaotic Systems. Phys. Rev. Letts. 64, 821–824 (1990)

[50] Liu, Z., Lai, Y.C., Hoppensteadt, F.C.: Pase clustering and transition to phase synchronization in a large number of coupled nonlinear oscillators. Phys. Rev. E 63, 055201 (2000)

[51] di Bernardo, M.: An adaptive approach to the control and synchronization of continuous-time chaotic systems. Int. J. of Bifuc. and Chaos 6, 557 (1996)
[52] Liu, Z., Shigang, C.: General method of synchronization. Phys. Rev. E 55, 199–204 (1997)
[53] Femat, R., Capistran-Tobías, J., Solís-Perales, G.: Laplace domain controllers for chaos control. Phys. Letts. A 252, 27–36 (1999)
[54] le Noble, W.J.: Highlights of organic chemistry. Marcel Dekker (1974)
[55] Bowong, S.: Stability analysis for the synchronization of chaotic systems with different order: application to secure communications. Phys. Lett. A 326, 102–113 (2004)

6 Remarks on Chaos Synchronizability and Synchronization

6.1 Complete Synchronizability

Synchronous behavior signifies that two or more systems operate or occur at the same time and space. Then, the synchronization problem is to induce the synchronous behavior in two or more dynamical systems. Thus, as a consequence, the synchronizability is defined as the property of dynamical systems such that the synchronous behavior can be induced on two or more dynamical systems. Such a problem can be understood as the interconnection of two or more chaotic systems in such manner that they behave in synchronous way. The synchronization can be performed by feedforward or feedback interconnections [1], [2]. In this sense synchronization problem is a control one; since synchronizability is related to the intrinsic properties of nonlinear systems to be synchronized, and such properties can be studied by exploiting the control theory concepts; controllability and observability.

There are two important problems related to chaos synchronization; one is the classification of the synchronization phenomena and the other is synchronizability of chaotic systems. The first one deals with the problem of classify the diverse phenomena in terms of the schemes and strategies used to achieved the synchronous behavior. For instance in the first problem Brown and Kocarev [3] proposed an intuitive definition for chaos synchronization of deterministic finite dimensional dynamical systems. Another approach is presented by Boccaletti *et.al.*[4], who gave a more general definition, including those which are not included in the Brown-Kocarev's work. On the other hand, Femat and Solís-Perales [5] showed some different forms of synchronization between third order chaotic systems as well as some definitions for these phenomena. Although, some important results are in literature about this part of the problem, less attention has been payed on the synchronizability problem as was introduced in the previous paragraph.

In this chapter we study the properties of chaotic systems to be synchronized due to its potential applications, for instance, in biological systems [6], [7], [8] or in secure communication [9], [10], [11]. The problem is the understanding of the mechanisms by

R. Femat & G. Solis-Perales: Robust Syn. of Chaotic Sys. Via Feedback, LNCIS 378, pp. 177–195, 2008.
springerlink.com

means of which the chaos synchronization is achieved, *i.e.*, synchronizability. Several approaches demonstrate that chaos synchronization can be achieved in different kinds, for instances complete synchronization, almost-synchronization, partial-state synchronization [5], phase synchronization [12], generalized synchronization [13], projective synchronization [14], etc. However, few results in the sense of synchronizability of chaotic systems are in literature [15], [16]. Then an actual question is: what conditions should the dynamical systems satisfy in order to they can be synchronized?, one result is presented by Udwadia and Raju [15], the authors studied conditions for synchronization between chaotic maps.

Another approach for synchronizability was reported by Josic [16]. The author reported sufficient and necessary conditions departing from the theory of invariant manifolds. The conditions are based on the existence of a diffeomorphism between the coupled system attractors, which is so called synchronization manifold. Using a modification of the Fenichel invariant manifolds theory and Lyapunov exponents, Josic found conditions for the stability of such a synchronization manifold. One restriction is that the manifold is r - normally hyperbolic, which implies that the flow contraction is r times exponentially stronger in the normal direction than in the interior of the synchronization manifold. Which means that the orbits of the driven system are attracted toward the synchronization manifold. To measure the attracting character of the synchronization manifold the Lyapunov exponents are used.

It is important to study the problem of synchronizability between dynamical chaotic systems in order to elucidate the topological and geometrical properties of the synchronous behavior and moreover to understand the interactions between systems which provide different phenomena [5]. Another important point about synchronizability is to study the effects of feedback or feedforward in the slave system, in other words, the study of the effects of couplings due to interconections between systems. The conditions are based on the properties of controllability and observability of nonlinear affine systems [17], [18], [19]. Since the chaos synchronization problem is interpreted as a control problem, controllability and observability conditions are applied to a dynamical system which represents the discrepancy between the master and slave system. The conditions provide a method to find an output function for the dynamical error system representing the discrepancy. This output function is such that the system is locally observable and controllable. However, a more detailed theory for synchronization, which is provided in Sec. 6.2, where a framework for synchronization is discussed.

6.1.1 The Chaos Synchronization Problem

The chaos synchronization consists in the tracking of the master system trajectories by the slave system. However, this problem can be restated as the stabilization of the orbits of a dynamical system which represents the discrepancy between systems. Such a dynamical system can be constructed as the difference between the master and slave system as in the following definition.

Definition 6.1 Let $\dot{x} = F_M(x)$ and $\dot{y} = F_S(y) + g(y)u$ be two chaotic systems in a manifold $\mathbf{M} \subset \mathbb{R}^n$ with F_M, F_S smooth vector fields, with output functions $s_M = h(x)$, $s_S = h(y)$ and $x, y \in \mathbb{R}^n$ and $g(y) \in \mathbb{R}^n$ is an smooth input vector. Let $w := x - y$ be the error between states and $y = x - w$, the dynamical error system can be written as $\dot{w} = F_M(x) - F_S(x - w) - g(x - w)u$, where $u \in \mathbb{R}$ is the control command.

In this section we have restricted our selves to synchronization error systems as in Definition 6.1 such that they can be written as a disturbed system $w = F_S(w)-g(w)u + \Psi(x,w)$, where $\Psi(x,w)$ represents the discrepancy term and is due to the differences between models; *i.e.*, to the nonlinear functions in both systems. Since the perturbation $\Psi(x,w)$ depends on the solution of the master system, the synchronization error system is extended in order to consider the flow of $x = F_M(x)$. That is, the synchronization error system is unidirectionally coupled with the master system. Therefore, from Definition 6.1, one can write an extended system to represent the synchronization error system as follows

$$\dot{x} = F_M(x) \tag{6.1}$$

$$\begin{aligned} \dot{w} &= F_M(x) - F_S(x - w) - g(x - w)u \\ Y_e &= h(x, w) \end{aligned} \tag{6.2}$$

where $Y_e = h(x) - h(x - w)$ is the output of the synchronization error system. System (6.1-2) can be written in affine form as: $X = F(X) + G(X)u$, where $X = [x,w]^T$, $F(X) = [F_M, F_M - F_S]^T$ and $G(X) = [0, -g]^T$.

In this sense one can consider that in system (6.1-2) $X = [x,w]^T \in \mathbb{R}^{2n}$, however, as we shall see bellow, the dimension of the synchronization manifold is at most n. Synchronizability of system (6.1-2) is studied from the case of Complete Synchronization (CS) [5], which means that every state on the slave system displays synchronous behavior with those corresponding to the master system. This is, subsystem (6.2) should be stabilized around the point $w^* = 0$ in order to achieved CS.

Definition 6.2. It is said that two chaotic systems are Completely Synchronized if the synchronization error $w \in \mathbb{R}^n$ remains into an arbitrarily small neighborhood containing the origin for all $t > t_o > 0$ and $(x(0),w(0))^T \in U \subset \mathbb{R}^{2n}$.

Subsystem (6.2) should be stabilized at origin by means of the control command u in face of the solution x of subsystem (6.1), therefore the problem reduces to find such a control command. The control command can be found by exploiting the geometrical control tools (*i.e.*, the properties of controllability and observability for nonlinear affine systems [17], [18], [19]. From such conditions the control law u can be designed such that subsystem (6.2) can be fully or partially linearized around the origin. Thus, the chaos synchronization problem can be formulated as the existence of a linearizing control law for system (6.1-2). The existence of the control law depends on the properties of controllability and observability of system (6.1-2), which provides the sufficient and necessary conditions for complete synchronizability of chaotic systems in affine form.

6.1.2 Synchronizability from Control of Chaotic Systems

We begin considering a definition for synchronizability, for complete synchronization between chaotic systems.

Definition 6.3. Synchronizability is an intrinsic property of dynamical systems such that we can induce a synchronous behavior by certain external force $u = u(x_M\text{-}x_S)$, *i.e.*, $\exists\ u = u(x_M\text{-}x_S) \in \mathbb{R}$ such that $|\ x(\mathrm{t}) - y(\mathrm{t})\ | \approx 0$, with $x, y \in \mathbb{R}^n$ and for all $\mathrm{t} > \mathrm{t}^* < \infty$ and any initial conditions $x_o = x(\mathrm{t} = 0)$ belonging to that manifold $\mathbf{M} \subset \mathbb{R}^{2n}$.

In terms of differential geometry the aim is to find a tangent space at the stabilization point $w = 0$, which can be determined by using the controllability and observability conditions. Note that in system (6.1-2) the input vector is $\mathrm{G} = [0, -\mathrm{g}]^T$ the 0 part correspond to the input vector in the master system. That is, the subsystem (6.1) is uncontrolled, unrestricted because it stands for the dynamics of the master system. The vector G can be calculated in such manner that it generates a tangent space with constant dimension at $(x,0)^T$, in other words, the dimension should be independent of the flow given by subsystem (6.1). On the other hand, the output function is defined as $Y_e = h(x,w)$ and its computation involves conditions for synchronizability of chaotic systems. The computation of the function $h(x,w)$ is such that the Lie derivatives of the output function along the system trajectories $\mathrm{X(t)} = [x(\mathrm{t}),w(\mathrm{t})]^T$ are independent around the region $\mathbf{E}_s = (x,0)^T$. This implies that system (6.1-2) is locally controllable and locally observable at the region $\mathbf{E}_S$.

Local Controllability for Complete Synchronizability

In order to deal with the problem of local controllability for chaos synchronizability, proof of the local accessibility is required. Recall that master and slave systems lie in a manifold, while the synchronization system is in a manifold $\mathfrak{M} \subset \mathbb{R}^{2n}$. Now, let $\mathbf{C} = \mathrm{span}\{\mathrm{X}(x,w)\ |\ \mathrm{X} \in \mathfrak{C},\ (x,w) \in \mathbb{R}^{2n}\}$ be the accessibility distribution of system (6.1-2), where $\mathfrak{C}$ is the accessibility algebra [18], [19] and X are linearly independent vector fields; **C** is a linear space in $\mathbb{R}^{2n}$ which is generated by $2n$ linearly independent vector fields and the accessibility algebra $\mathfrak{C}$ is a set of vector fields which defines all the possible directions from where the system is accessible. Thus, from system (6.1-2), we can consider that the dimension of the accessibility distribution is d for all $(x,w) \in \mathbb{R}^{2n}$ with the properties: (i) **C** is involutive, (ii) contains the distribution span{G} and (iii) is invariant under the vector fields $\mathrm{F}(x,w)$ and $\mathrm{G}(x,w)$. Therefore, for every point $(x,w)^o$ a neighbourhood U^o of $(x,w)^o$ can be found and a local coordinate transformation such that the accessible points at time $\mathrm{T} < \infty$ along the trajectories in U^o remain in U^o.

Lemma 6.4. [18], [19] Let $\mathbf{C}(x,w)$ be the accessibility distribution with constant dimension d at $w = 0$ then system (6.1-2) is locally accessible (locally controllable) and the controllable space has dimension d.

Therefore, with Lemma 6.4 in mind, we can find a local coordinate transformation such that system (6.1-2) can be partially or completely transformed into a linear controllable system around $(x,0)^T$. To this end, we need to find the accessibility distribution for system (6.1-2). We first consider the following definition for Lie brackets and Lie derivative.

Definition 6.5. [18], [19] Let X and Y be two vector fields in a manifold **M**, the vector field $[X,Y](z) = X_z(Y(z)) - Y_z(X(z))$ is called the Lie bracket of X and Y. The Lie derivative of a real-valued function f along the vector field X is defined as $L_X f(z) = \langle df, X \rangle$.

Without lost of generality we can consider vector fields X_i with the property $[X_i, X_j] \neq 0$ with $i \neq j$ and $i,j = 1,2,...,n$ into a neighborhood of the point x^o and generate a tangent space at x^o. In other words, for each point $x \in \mathbf{M}$ there exists a neighborhood U of x and a set of smooth vector fields X_i, such that, the distribution $\mathbf{C}(x) = \text{span}\{X_i(x)\}$ with $i = 1,2,...,I$, has constant dimension, I is some integer index. Another important fact is that a distribution **C** is involutive if the Lie bracket $[X_i,X_j] \in \mathbf{C}$ with X_i and X_j are in **C**. Now we can find the vector fields that generate a tangent space from vector fields F and G of system (6.1-2).

The accessibility distribution $\mathbf{C}(x,w)$ can be determined as $\mathbf{C}_k(x,w) = \text{span}\{\mathbf{ad}_F^{k-1}\mathbf{G}\}$ where $\mathbf{ad}_F^{k-1}\mathbf{G} = [F,[F,[...,[F,G],...,]]]$ for $k = 1...2n$. Now recall that $F = [F_M(x), F_M(x) - F_S(x-w)]^T$ and $G = [0,-g]^T$ then the Lie brackets are given as follows

$$ad_F G = \begin{bmatrix} 0 \\ -F_M(x)\dfrac{\partial g}{\partial x} - (F_M(x) - F_S(x-w))\dfrac{\partial g}{\partial w} + \dfrac{\partial F_S(x-w)}{\partial w} \end{bmatrix} \tag{6.3}$$

now we take $\Psi(x,w) = -F_M(x)\partial g/\partial x - (F_M(x) - F_S(x - w))\partial g/\partial w + \partial F_S(x - w)/\partial w$ and the Lie bracket $\mathbf{ad}_F^2\mathbf{G}$

$$ad_F^2 G = \begin{bmatrix} 0 \\ F_M(x)\dfrac{\partial \Psi(x,w)}{\partial x} + (F_M(x) - F_S(x-w))\dfrac{\partial \Psi(x,w)}{\partial w} + \Psi(x,w)\dfrac{\partial F_S(x-w)}{\partial w} \end{bmatrix} \tag{6.4}$$

from this vector fields it can be find the accessibility distribution $\mathbf{C}_k(x,w)$. This distribution is generated by k linearly independent vector fields that span a tangent space of dimension $d \leq n$, this because the subsystems (6.1) is unidirectionally coupled with subsystem (6.2). Since system (6.1-2) is composed by two subsystems one given by the master and the other by the difference between master and slave, the tangent space locally generated at $(x,0)^T$ has dimension $d \leq n$. This means, that it is not required to calculate $2n$ vector fields to generate the tangent space, moreover the accessibility space is of dimension $d \leq n$.

There are two restrictions concerning the controllability for synchronizability of master/salve systems: (i) the constant dimension $\text{Dim}(\mathbf{C}_d(x,0)) = d$ and (ii) the distribution $\mathbf{C}_{d-1}$ should be involutive. In case of the dimension of the tangent space be $d < n$ implies the stability analysis of a subsystem of dimension $2n - d$. This analysis is the so called internal dynamics and should be stable or bounded to achieve stability on the linearized system [18], [19]. Finally with these results the controllability for complete synchronizability is solved.

Proposition 6.6. Suppose that the internal dynamics is stable and assume that the accessibility distribution **C** has constant dimension d (*i.e.*, **C** is involutive) in a neigborhood U^o of $(x,0)^T$, then system (6.1-2) is locally controllable.

Local Observability for Complete Synchronizability

In the problem of local controllability the output function of system (6.1-2) is not considered. The output function $Y_e(x,w) = h(x,w)$, where h is a real-valued function of the master and error system is related to the observability of the synchronization system. In order to system (6.1-2) satisfies local observability the output function should be determined. From the geometrical control theory let us consider the definition of the observation space.

Definition 6.7. Consider the affine system (6.1-2) and the output function $Y_e(x,w) = h(x,w)$. The observation space $\mathcal{O}$ is the linear space of functions in **M** that contains $h(x,w)$ and all the Lie derivatives of the output function along the system trajectories.

In the same way than controllability, here the observation space is determined departing from the vector field F in system (6.1-2) and we assume that there exists an output function $h(x,w)$. Without lost of generality, we can consider the vector fields from the accessibility distribution to define the observability codistribution $d\mathcal{O}$. Such codistribution is involutive and if $\mathrm{Dim}(d\mathcal{O}) = n$ then system(6.1-2) is locally observable at $(x,0)^T$. On the other hand, if $\mathrm{Dim}(d\mathcal{O}) = d < n$, by Frobenius Theorem there exists a local coordinate transformation such that system (6.1-2) is locally observable. Let us consider the following well known definition.

Definition 6.8. [18], [19] Consider an affine system $\dot{x} = f(x) + g(x)u$, with an output function $y = h(x)$. It is said that the system has relative degree ρ at x^o if

(i) $L_g L_f^k h(x) = 0$ for all x in a neighborhood of x^o and $k < \rho - 1$
(ii) $L_g L_f^{\rho-1} h(x^o) \neq 0$

Now consider that system (6.1-2) has relative degree ρ at $(x,0)^T$, the Lie derivatives of the output function along the system trajectories are given by $h(x)$, $L_F h(x)$, $L_F^2 h(x),\ldots,L_F^{\rho-1} h(x)$, and the covector fields $dh(x)$, $dL_F h(x)$, $dL_F^2 h(x),\ldots,dL_F^{\rho-1} h(x)$, which are independent in the neighborhood U of $(x,0)$ [18], [19]. In this way, the Lie derivatives of $h(x,w)$ can be used to define a local coordinate transformation at $(x,0)^T$, recall that $w = 0$ *means complete synchronization*. Such a transformation $z = \Phi(x,w)$ is defined as $\phi_1 = h(x)$, $\phi_2 = L_F h(x),\ldots$, $\phi_\rho = L_F^{\rho-1} h(x)$ and the $2n - \rho$ complementary functions are determined in such a way that the Jacobian matrix of $z = [\phi_1\ \phi_2,\ldots, \phi_{\rho,\ldots}, \phi_{2n}]^T$ be nonsingular at $(x,0)$ and $L_g \phi_i(x,w) = 0$, with $i = \rho + 1,\ldots,2n$ and (x,w) in a neighborhood of $(x,0)$.

Proposition 6.9. Suppose that system (6.1-2) has relative degree ρ at $(x,0)$ and the codistribution $\ker(\mathrm{span}\{dh(x,0), dL_F h(x,0), dL_F^2 h(x,0),\ldots,dL_F^{\rho-1} h(x,0)\})$ has dimension ρ, then system (6.1-2) is locally observable at $(x,0)$.

6.1.3 Complete Synchronizability of Chaotic Systems

Once that controllability and observability conditions for complete synchronizability hold for system (6.1-2), we have that synchronization error system is linearizable by feedback. Such a linearization is carried out by an invertible coordinates transformation $z = \Phi(x,w)$ and a control command $u = u(x_M - x_S)$ which can be calculated from the transformed system. Finally, from the theory of linear systems the synchronization error system (6.1-2) can be stabilized at $(x,0)$ [20].Subsystem (6.2) is stabilized at $w = 0$ by state feedback if it is locally controllable and locally observable and, as a consequence, dynamical systems comprised in Definition 6.1 are synchronizables. Condition for complete chaos synchronizability can be reduced to the following Theorem which comprises the above mentioned stabilizing conditions.

Theorem 6.10. (Complete Synchronizability) Consider system (6.1-2). Suppose that there exist $2n - \rho$ functions $\phi_i(x,w)$ such that $L_g\phi_i(x,w) = 0$, $i = 2n - \rho...2n$. System (6.1-2) is feedback linearizable at $(x,0)$ if and only if there exists a function $h(x,w)$ such that

(i) $<dh, ad_F^{k-1}G>(x,w) = 0$ for $k = 1,..., \rho - 1$; $\rho > 1$ and (x,w) in a neighborhood U of $(x,0)$

(ii) $<dh, ad_F^{i}G>(x,0) \neq 0$ for $i = \rho....n$, at $(x,0)$

where $\rho = d$ stands for the dimension of the tangent space.

Proof. First suppose that (i) and (ii) hold, then the functions $h(x,w)$, $L_Fh(x,w)$,...,$L_F^dh(x,w)$, are independent around $(x,0)$ and from (i) we have that the distribution $C_i \subset \ker(\text{span}\{dh(x,0), dL_Fh(x,0), dL_F^2h(x,0),...,dL_F^{d-1-i}h(x,0)\})$ has dimension i and, without lost of generality, the dimension of $\text{span}\{dh(x,0), dL_Fh(x,0), dL_F^2h(x,0),...,dL_F^{d-1-k}h(x,0)\} = d - k$ where $k = 1,...,d - 1$ and $d > 1$ which are all involutive and constant dimension, therefore from Propositions 6.6 and 6.9, system (6.1-2) is feedback linearizable at $(x,0)$.

Conversely, suppose that system (6.1-2) is feedback linearizable, it implies that there exists a local coordinate transformation $\Phi(x,w)$ which is invertible, besides, there exists a linear subspace of dimension $d = \rho$ at $(x,0)$ and as a consequence the existence of a function $h(x,w)$. Such linear subspace is generated by exactly ρ vector fields which are linearly independent around $(x,0)$, the dimension of $C_\rho(x,0)$ is constant, on the other hand, the covector fields $dL_F^kh(x,w)$ are linearly independent therefore, (i) and (ii) hold. □

Remark 6.11. Theorem 6.10 provides sufficient and necessary conditions for chaos complete synchronizability of same order systems. Moreover, conditions (i) and (ii) can be used to calculate an output function such that the dynamical error system (6.1-2) is locally observable at $(x,0)$. In the same way it can be calculated an input vector for a given output function such that system (6.1-2) is locally controllable.

Note that in Theorem 6.10 the case of dimension $d = 1$ for the accessibility distribution is not considered, because there is only one direction in which the system is accessible, this direction is given by the input vector field $G(w)$ and the output function can be determined from condition (ii). Then we have a Corollary for synchronizability of two same order chaotic systems.

Corollary 6.12. Two chaotic systems with same order and similar model are completely synchronizable if and only if the dynamical error system (system (6.1-2)) is feedback linearizable at $(x,0)$.

Now let us consider the output function $y = h(x,w)$ and functions $L_F^{i-1}h(x,w)$ define the coordinate transformation $\Phi(x,w)$, where $i = 0,1,...,\rho$, and in order to find the transformed system we consider $\dot{z}_k = L_F^kh(x,w) = z_{k+1}$ where $k = 1,2,...,\rho - 1$ and for

$k = \rho$ $\dot{z}_\rho = L_F^\rho h(x,w) + L_g L_F^{\rho-1} h(x,w)u$, and the transformed system can be written as follows

$$\begin{aligned}
\dot{z}_1 &= z_2 \\
\dot{z}_2 &= z_3 \\
&\vdots \\
\dot{z}_{\rho-1} &= z_\rho \\
\dot{z}_\rho &= \beta(\xi,\eta) + \alpha(\xi,\eta)u \\
\dot{\eta} &= \Theta(\xi,\eta)
\end{aligned} \tag{6.5}$$

where $\xi = [z_1, z_2, ..., z_\rho]$, $\eta = [z_{\rho+1}, ..., z_{2n}]$ and $\dot{\eta} = \Theta(\xi,\eta)$ represents the internal dynamics of the transformed system. System (6.5) has two parts, the first one represents the linearized system and the second one is in general a nonlinear system which represents the unobservable and uncontrollable dynamics, since in this subsystem there is no influence of the control command and no state can be observed. The control command that stabilizes system (6.5) is

$$u = \frac{1}{\alpha(\xi,\eta)}(-\beta(\xi,\eta) + \kappa_i(z_i - z_i^*)) \tag{6.6}$$

where κ_i with $i = 1,2,...,\rho$ are the control gains and are chosen in such a way that the closed-loop subsystem $\dot{\mathbf{z}}$ converges to the origin and $\mathbf{z}_i^*$'s are the coordinates of the stabilization point and the functions $\alpha(\xi,\eta) = L_g L_F^{\rho-1} h(x,w)$ and $\beta(\xi,\eta) = L_F^\rho \boldsymbol{h}(\boldsymbol{x},\boldsymbol{w})$. A necessary and sufficient condition to stabilize system (6.5) is that the internal dynamics be stable or bounded. To determine stability of the internal dynamics, the zero dynamics is analysed by setting the states z's to zero. In this way, the closed-loop system with $\dot{z}_1(t) = \dot{z}_2(t) = ... = \dot{z}_\rho(t) = 0$, is $\dot{\eta} = \Theta(0,\eta)$ which is the zero dynamics and should be stable or bounded for all η.

6.1.4 Illustrative Example

In order to show the results in previous sections, we chose two Lorenz systems. For simplicity we consider systems with same parameter values

$$\begin{aligned}
\dot{x}_1 &= \sigma(x_2 - x_1) \\
\dot{x}_2 &= \rho x_1 - x_2 - x_1 x_3 \\
\dot{x}_3 &= x_1 x_2 - \beta x_3
\end{aligned} \tag{6.7}$$

$$\begin{aligned}
\dot{y}_1 &= \sigma(y_2 - y_1) + g_1(y)u \\
\dot{y}_2 &= \rho y_1 - y_2 - y_1 y_3 + g_2(y)u \\
\dot{y}_3 &= y_1 y_2 - \beta y_3 + g_3(y)u
\end{aligned} \tag{6.8}$$

where the parameters $\sigma = 10$, $\rho = 28$ and $\beta = 8/3$. Functions g_i with $i = 1,2,3$ are the corresponding elements in the input vector of the slave system and defines how the control command modifies the vector field. This functions can be chosen in such way that the tangent space has complete and constant dimension, in this example we chose this functions as constants. The control command should be designed such that system (6.8) be synchronous in all states (complete synchronization) with system (6.7). On the other hand, note that there is no defined an output function in any system, this because an appropriate output function shall be calculated.

From Definition 6.1 the synchronization error system can be written as a system $\Sigma \in \mathbb{R}^6$ as follows

$$\begin{aligned}
\dot{x}_1 &= \sigma(x_2 - x_1) \\
\dot{x}_2 &= \rho x_1 - x_2 - x_1 x_3 \\
\dot{x}_3 &= x_1 x_2 - \beta x_3 \\
\dot{w}_1 &= \sigma(w_2 - w_1) - g_1(x - w)u \\
\dot{w}_2 &= \rho w_1 - w_2 + w_1 w_3 - w_1 x_3 - x_1 w_3 - g_2(x - w)u \\
\dot{w}_3 &= w_1 w_2 - \beta w_3 + w_1 x_2 + x_1 w_2 - g_3(x - w)u
\end{aligned} \tag{6.9}$$

note that subsystem w is perturbed by the flow of system (6.7), this is w = F(w) + $\Psi(w,x)$ + G(w,x)u, where F, G and $\Psi \in \mathbb{R}^6$ are vector fields and $\Psi(x,w)$ is a perturbation due to the nonlinearities in both systems.

First we calculate the distribution at (x,0) for system (6.9) and we look for a vector G such that the tangent space has dimension d = 3 (see Remark 6.3). Recall that we should calculate $\mathbf{C}_k(x,w)$ = span{adkF^{-1} G} with k = 1,...,d. Let us consider G = $[0\ 0\ 0\ g_1\ g_2\ g_3]^T$ with g_i constants and calculate the corresponding distributions. The distribution $\mathbf{C}_1(x,w)$ = span{G}, $\mathbf{C}_2(x,w)$ = span{G, ad_FG} where ad_FG is given by

$$ad_F G = \begin{bmatrix} 0 \\ 0 \\ 0 \\ \sigma(g_1 - g_2) \\ g_1(w_3 + x_3 - \rho) + g_2 + (w_1 + x_1)g_3 \\ - g_1(w_2 + x_2) - (w_1 + x_1)g_2 + \beta g_3 \end{bmatrix}$$

and $\mathbf{C}_3(x,w)$ = span{G, ad_FG,adF2G} with adF2G given by

$$ad_F^2 G = \begin{bmatrix} 0 \\ 0 \\ 0 \\ \sigma^2(g_1 - g_2) - \sigma((w_3 + x_3 - \rho)g_1 + g_2 + (w_1 + x_1)g_3) \\ g_1[\sigma(\rho - w_3 - x_3) + (w_1 w_2 - \beta w_3 + w_1 x_2 + x_1 w_2) - (\rho - w_3 - x_3)(\sigma + 1) - (w_1 + x_1)(w_2 + x_2)] + \\ g_2[\sigma(\rho - w_3 - x_3) - (w_1 + x_1)^2 + 1] \\ - g_1[(\rho x_1 - x_2 - x_1 x_3) + (\rho w_1 - w_2 - w_1 w_3 - w_1 x_3 - x_1 w_3) + (w_2 + x_2)(\sigma - \beta) - (w_1 + x_1)(\rho - w_3 - x_3)] - \\ g_2[\sigma(x_2 - x_1) + \sigma(w_2 - w_1) - \sigma(w_2 + x_2) + (w_1 + x_1)(1 + \beta)] - g_3[(w_1 + x_1)^2 - \beta^2] \end{bmatrix}$$

Note that the dimension of $\mathbf{C}_3(x,0)$ is $d \leq 3$. Now, we have from ad_FG(x,0) that the last two components are different from zero if $g_1 = 0$, $g_2 > g_3 | x_1 |$ and $g_3 > g_2/\beta | x_1 |$. However, this condition depends on the flow x_1(t) which is chaotic and cannot be modified, as a consequence $\mathbf{C}_3(x,0)$ has variable dimension. On the other hand, it is not possible to satisfy conditions (i) and (ii) in Theorem 6.10. From this result we find that two Lorenz systems cannot generate a tangent space of dimension d = 3 with an input vector of constant elements. With this in mind we proceed to determine a tangent space of constant dimension $d = 2$. Without lost of generality, we can consider $G = [0\ 0\ 0\ 0\ -1\ 0]^T$ and $\mathbf{C}_3(x,w) = \text{span}\{G, ad_F G, \mathbf{ad}_F^2 G\}$ is given as follows

$$C_3(x,0) = span\left\{\begin{bmatrix} 0 & 0 & 0 \\ 0 & 0 & 0 \\ 0 & 0 & 0 \\ 0 & \sigma & \sigma^2 + \sigma \\ -1 & -1 & (+1+(\rho - x_3)\sigma - x_1^2) \\ 0 & x_1 & x_1(-\sigma + 1 + \beta) \end{bmatrix}\right\}$$

Note that this distribution generates a space of dimension $d = 3$ if $x_1(-\sigma + 1 + \beta) \neq 0$, again we cannot restrict the flow $x_1(t)$ then system (6.9) is not locally controllable, however, $\text{Dim}(\mathbf{C}_3(x,0)) = 2$ for all $x(t)$. Now from Theorem 6.10 we can calculate the output function $h(w)$ such that

$$\sigma \frac{\partial h}{\partial w_2} = 0 \tag{6.10}$$

$$\sigma \frac{\partial h}{\partial w_1} - \frac{\partial h}{\partial w_2} + x_1 \frac{\partial h}{\partial w_3} \neq 0 \tag{6.11}$$

a possible function which satisfy (6.10) and (6.11) is $h(x,w) = w_1$. Once we have an output function the relative degree is calculated and for this case is $d = \rho = 2 = \text{Dim}(\mathbf{C}_3(x,0))$ for all $x \in \mathbb{R}^3$ and we have that $L_g h(w,x) = 0$, $L_g L_F h(x,w) = -\sigma$ then the transformation is given as $z_1 = h(x,w) = w_1$, $z_2 = L_F h(x,w) = \sigma(w_2 - w_1)$ and to complete the transformation, it is required to find four functions ϕ such that $L_g\phi = 0$ and we have the transformation

$$\begin{aligned} z_1 &= w_1 \\ z_2 &= \sigma(w_2 - w_1) \\ z_3 &= w_3 \\ z_4 &= x_1 \\ z_5 &= x_2 \\ z_6 &= x_3 \end{aligned} \tag{6.12}$$

which is an invertible transformation around $(x,0)$. The transformed system is feedback linearizable, which means that there exists a control command $u(x,w)$ such that the trajectories of system (6.8) are close to those of system (6.7). Then the transformed system can be written as follows

$$\begin{aligned} \dot{z}_1 &= z_2 \\ \dot{z}_2 &= L_F^2 h(x,w) + L_G L_F h(x,w) u \\ \dot{z}_3 &= \dot{w}_3 \\ \dot{z}_4 &= \dot{x}_{1,M} \\ \dot{z}_5 &= \dot{x}_{2,M} \\ \dot{z}_6 &= \dot{x}_{3,M} \end{aligned} \tag{6.13}$$

where the control command can be defined as (6.6) with $\alpha(\Phi^{-1}(z)) = L_g L_F h(x,w) = -\sigma$ and $\beta(\Phi^{-1}(z)) = L_F^2 h(x,w) = -\sigma^2(w_2 - w_1) + \sigma(\rho w_1 - w_2 + w_1 w_3 - w_1 x_3 - x_1 w_3)$. We can chose control gains κ's such that the subsystem $(z_1, z_2)^T$ is stable at $(x,0)$. The internal dynamics given by $(z_3, z_4, z_5, z_6)^T$ is bounded since $x_{i,M}$, i = 1,2,3 corresponds to the dynamics of the master system which is chaotic and is contained into the attraction region **E**, and the state $\dot{z}_3 = \dot{w}_3 = (z_1 + z_4)(z_2 + \sigma z_1)/\sigma - \beta z_3 + z_1 z_2$ which for the zero dynamic is only $\dot{z}_3 = -\beta z_3$, is stable.

Therefore, geometrical properties of chaotic systems can be considered to obtain conditions for chaos complete synchronizability. The key properties for complete synchronizability are local controllability and local observability for nonlinear affine systems. Such conditions provide a design method for the output function of the dynamical error system. This output function is such that both master and slave systems can be synchronized by state feedback. Thus, Theorem 6.10 can be used to determine if two chaotic systems can be completely synchronized (complete synchronizability). Then in order to have a better understood of chaos synchronization more attention on the so called synchronization function $\lambda = h(x)$, is detailed in the next section. This synchronization function is defines the relationship between both systems, and it provides information on the synchronization in general terms.

6.2 General Framework of Chaotic Synchronization Based on Lie Algebra

Synchronous behavior can be found in diverse dynamical systems (as, to mention some, biological [21] or communication [22]) and its study is theoretically interesting and technologically important. Nevertheless, there are diverse synchronization phenomena related to the chaotic systems [5]. Indeed, synchronization has been induced in strictly different oscillators [1], [23] and systems with different order [24]. Such fact increases the complexity of the chaotic synchronization meaning. In the nonlinear science, the definition of the synchronous behavior means that the trajectories of two or more dynamical systems evolve, in some sense, close along their trajectories. As a consequence of the diversity of the synchronization phenomena, the actual discussion point is the unification toward a general definition of the synchronous behavior in chaotic systems [3], [4]. The idea is that the different kinds of synchronization can be captured by a formalism [3] by searching the existence of a differmorphism between attractors of the coupled systems [16] whose properties involve a time-invariant synchronization manifold (some authors called it "the synchronization function" [4]).

The efforts have been focussed to the analysis of time-invariant manifolds applied to synchronization [4], [16], [25]. Two basic approaches have been exploited. On the one hand, chaotic synchronization has been interpreted as the prediction of the chaotic system, *i.e.*, observability approach. In this sense, the reconstruction of the drive system attractor from the response system is interpreted as an observer [25]. An interesting point about observability of the synchronization systems is that differential geometry allows to find an invariant space under vector fields where the attractor can be reconstructed (see Chapter 1 in [18]). Synchronization of chaotic systems, on the other hand, has been also studied from the measurable variables (system output) [1], [3], [24].

In such a case, the synchronization is understood as a control (stabilization) problem. In other words, to compute the controller such that the difference between trajectories of the slave system $x_s(t)$ remains close to those of the master system $x_M(t)$. That is, the point is to find the invariant space such that the origin of the synchronization error system $\|x_M(t) - x_S(t)\| = 0$ can be stabilized. Both observability and controlability of nonlinear systems are included in the geometrical control theory [18], [26].

In this Section, the goal is to provide some remarks on the diffeomorphism between chaotic attractors by exploiting the Lie algebra of chaotic systems [18], [19] to compute the synchronization function. In this manner, we discuss both observation and stabilization approaches. That is, we are interested in the Lie-based geometric properties of the class of dynamical systems given by $x = \tau_M(x_M) - \tau_S(x_S) - g(x)u$, where $x \in \mathbb{R}^n$ is, by definition, $x := x_M - x_S$ and stands for the state vector of the synchronization error system, $\tau_M : \mathbb{R}^n \rightarrow \mathbb{R}^n$, $\tau_S : \mathbb{R}^n \rightarrow \mathbb{R}^n$ and $g : \mathbb{R}^n \rightarrow \mathbb{R}^n$ are smooth vector fields. The product $g(x)u$ is related to the synchronization command. The scalar function $u = u(x)$ can be computed from the construction of accessibility spaces [27]. Since we are interested in the output $y = \lambda(x)$ of the above-mentioned dynamical system, synchronization is formuled in terms of geometrical properties of the system along the vector fields $\tau(x,x_M) = \tau_M(x_M) - \tau_S(x_S)$ and $g(x)$. Thus the question is: what are the geometrical properties of any synchronization error system such that the scalar function $u = u(x)$ guarantees the existence of the synchronization function $y = \lambda(x)$? This is addressed via Lie algebra of the vector field related to the synchronization error system.

6.2.1 Lie-Based Geometry of Nonlinear Systems

The idea behind the Lie-based geometry is to find the coordinate decompositions of the synchronization error system. Thus, an invariant space (called distribution) can be constructed and analyzed. Such a space is spanned by the vector fields of the synchronization error system and the Lie Brackets between them. In seek of completeness, this section contains a brief description of the decompositions.

Lie Brackets and Distribution Notion

The starting point is the properties of the Lie Brackets. A Lie bracket is a kind of differential operation on a dynamical systems along its vector fields. Such an operation will be denoted by $[\tau_1,\tau_2]$ for the vector fields τ_1,τ_2: $\mathbb{R}^n \rightarrow \mathbb{R}^n$ and is defined by

$$[\tau_1, \tau_2](x) = \frac{\partial \tau_2}{\partial x}\tau_1(x) - \frac{\partial \tau_1}{\partial x}\tau_2(x) \tag{6.14}$$

at each x in the subset $U \subset \mathbb{R}^n$, where

$$\frac{\partial \tau_i}{\partial x} = \begin{bmatrix} \frac{\partial \tau_{i,1}}{\partial x_1} & \frac{\partial \tau_{i,1}}{\partial x_2} & \cdots & \frac{\partial \tau_{i,1}}{\partial x_n} \\ \frac{\partial \tau_{i,2}}{\partial x_1} & \frac{\partial \tau_{i,2}}{\partial x_2} & \cdots & \frac{\partial \tau_{i,2}}{\partial x_n} \\ \vdots & \vdots & \ddots & \vdots \\ \frac{\partial \tau_{i,n}}{\partial x_1} & \frac{\partial \tau_{i,n}}{\partial x_2} & \cdots & \frac{\partial \tau_{i,n}}{\partial x_n} \end{bmatrix} \tag{6.15}$$

stands for the Jacobian matrices of the mappings τ_i, i = 1,2. The differential operation (6.14) of a vector field τ_2 with the same vector namely τ_i, is given by $[\tau_1[...\tau_1[\tau_1,\tau_2]]...]$. In sake for simplicity in notation, the recursive *k-th* Lie bracket of the vector field τ_2 along the vector field τ_1 valuated at the point $x \in \mathbb{R}^n$, is denoted, by the expression $ad^k_{\tau 1}\tau_2 = [\tau_1, ad^{k-1}_{\tau 1}](x)$ for any k≥1 and $ado_1\ \tau_2(x) = \tau_2(x)$ [18], [19]. The Lie bracket between the vector fields τ_1 and τ_2 has the following properties:

- Is bilinear over $\mathbb{R}$, that is, if $\theta_1,\theta_2,\tau_1$and τ_2 are vector fields, $\pi_1$and π_2 are real numbers, then (i) $[\pi_1\tau_1+\pi_2\tau_2,\theta_1] = \pi_1[\tau_1,\theta_1] + \pi_2[\tau_2,\theta_1]$ and (ii) $[\tau_1,\pi_1\theta_1 + \pi_2\theta_2] = \pi_1[\tau_1,\theta_1] + \pi_2[\tau_2,\theta_2]$.
- Is skew symmetric $[\tau_1,\tau_2] = -[\tau_2,\tau_1]$.

Now, suppose that on a open set *U* there are *d* vector fields such that, for any given point *x* in *U*, the vector $\tau_i(x)$ span a vector space. Let us denote the assignment in a vector space to each point *x* of *U* as $\Delta(x) = \text{span}\{\tau_1(x)...\tau_d(x)\}$; Δ is called distribution. In order to illustrate such a notion, let us define F as a matrix having *n* columns whose entries are smooth functions of any variable, namely *x*. The columns of the matrix F can be interpreted as vector fields. Hence, such a matrix identifies the distribution spanned by its columns and can be valued by $\Delta(x) = \text{Im}(F(x))$, where Im(F(*x*)) denotes the image of the matrix F at any *x* in *U*. Of course, the dimension of the distribution can be associated to the rank of the matrix F at the point $x \in U$. Finally, let $\mathring{x}$ be in any open set *U*, it is said that $\mathring{x}$ is a regular point of a distribution Δ if there exists a neighborhood $\mathring{U}$ of $\mathring{x}$ such that the distribution Δ is nonsingular at any *x* in $\mathring{U}$. Thus, if Δ_1 and Δ_2 are distributions, then:

- The sum $\Delta_1+\Delta_2$ is defined by taking the sum of the subspaces $\Delta_1(x)$ and $\Delta_2(x)$, namely $(\Delta_1+\Delta_2)(x) = \Delta_1(x)+\Delta_2(x)$,
- The intersection $\Delta_1\cap\Delta_2$ is defined as $(\Delta_1\cap\Delta_2)(x) = \Delta_1(x)\cap\Delta_2(x)$.

Thus with these preliminary information on characteristics of distributions, we can proceed to illustrate some remarkable facts in terms of distribution, involutibity and integrability for vector fields.

6.2.2 Involutive Distributions and Flows

Definition 6.13. Let τ_1 and τ_2 two vector fields belonging to the distribution Δ. It is said that the distribution Δ is involutive if the Lie bracket $[\tau_1,\tau_2]$ is a vector field belonging to the distribution Δ.

Example 6.14. Let π_i,μ_i, i =1,2,3 be real positive constants and consider two vector fields τ_1 and τ_2 in $\mathbb{R}^3$ given by $\tau_1(x) = (\pi_1(x_2 - x_1),\pi_2x_1 - x_2 - x_1x_3, - \pi_3x_3 + x_1x_2)^T$ and $\tau_2(x) = (\mu_1(x_2 - x_1),\mu_2x_1 - x_2 - x_1x_3,-\mu_3x_3+ x_1x_2)^T$. Note that $\tau_1(x)$ and $\tau_2(x)$ define the vector fields of two Lorenz equations with parameters π_i and μ_i, respectively. In addition, *dim*(span $\{\tau_1(x), \tau_2(x)\}$) = 2 at any point $x \neq x_2$ and $x_3 \neq \pi_2$ far away the origin and for all parameters values $\pi_i,\mu_i \neq 0$. Thus, since the Lie bracket belongs to the distribution $\Delta(x)$ for any parameters values $\pi_i = \mu_i \neq 0$, hence, at any point far away the origin with $x_1 \neq x_2$ and $x_3 \neq \pi_2$, the distribution $\Delta(x)$ is involutive. The condition $\mu_i = \pi_i$,i = 1,2,3 means that both Lorenz equations are identical.

$$[\tau_1,\tau_2](x)=\begin{pmatrix}(\mu_1\pi_2-\pi_1\mu_2)x_1+(\pi_1-\mu_1)x_2+(\pi_1-\mu_1)x_1x_3\\(\mu_1\pi_2+\mu_2-\pi_1\mu_2-\pi_2)x_1+(\pi_1\mu_2-\mu_1\pi_2)x_2+(\pi_1-\mu_1+\pi_3-\mu_3)x_1x_3+(\mu_1-\pi_1)x_3x_2\\(\mu_1-\pi_1)x_2x_1+(\pi_2-\mu_2)x_1^2+(\pi_1-\mu_1)x_2^2+(\pi_3-\mu_3)x_1x_2\end{pmatrix}$$

Definition 6.15. A distribution Δ, defined on an open set U, is no singular if there exists an integer d such that $dim(\Delta(x)) = d$ for all x in U. Otherwise, the distribution is singular.

Definition 6.16. A point $\mathring{x} \in U$ is called regular point of a distribution $\Delta(x)$ if there exists a neighborhood $\mathring{U} \subset U$ of $\mathring{x}$ with the property that $\Delta(x^{o})$ is non singular on $\mathring{U}$.

Example 6.17. Let us consider the vector fields $\tau_1(x) = (f_1(x),f_2(x))^T$ and $\tau_2(x) = (0,1)^T$, where $f_1(x) = x_2$ and $f_2(x) = -\pi_1x_2 + x_1 - x31$ are analytical functions of $x \in \mathbb{R}^2$ whose parameter $\pi \in \mathbb{R}^p$ are no null. Note that, in this case, the vector field $\tau(x)$ represents the second-order oscillators representing the Duffing equation. Now, the distribution $\Delta(x) = \text{span}\ \{\tau_1(x),\tau_2(x)\}$ has dimension $d = 2$ on $U = \{x \in \mathbb{R}^2 : x_2 \neq 0\}$ and dimension $d = 1$ on the set $\Theta = \{x \in \mathbb{R}^2 : x_2 = 0\}$. Thus, the distribution $\Delta = \text{span}\ \{\tau_1,\tau_2\}$ is singular at the origin; however, it is regular elsewhere. The point $\tilde{x}$ is called no regular point or point of singularity. Note that the distribution Δ is involutive on U; however, Δ is not involutive at $\Theta \subset \mathbb{R}^2$.

An involutive distribution is directly related to the flow of dynamical systems by the Frobenius Theorem [18]. Which states that *a no singular distribution Δ is integrable if and only if it is involutive.* This implies that if a distribution Δ satisfies this theorem, there exist a local coordinate chart.

Example 6.18. Consider the following distribution defined on $\mathbb{R}^2$, $\Delta = \text{span}\ \{\tau_1,\tau_2\}$, where

$$\tau_1(x)=\begin{pmatrix}x_1^2\\-1\end{pmatrix};\qquad \tau_2(x)=\begin{pmatrix}x_2+x_1(1-x_2^2-x_1^2)\\-x_1+x_2(1-x_2^2-x_1^2)\end{pmatrix}$$

therefore $\dim(\Delta(x)) = 2$ for any x in $U = \{x \in \mathbb{R}^2\text{: } x_2 = x_1,\ 0 < |x_1| < 2((5)^{1/2} - 1)^{1/2}\}$. Note that the distribution $\Delta(x)$ is involutive on U. The flow of $\tau_1(x)$ and $\tau_2(x)$ are related to the solution of $x = \tau_i(x)$, $i = 1,2$. Thus, by defining $\xi_0 = (\xi_{1,0},\xi_{2,0}) \in \mathbb{R}^2$ as the initial conditions $\tau_0 = 0$. About $\tau_1(x)$, the equation $x = \tau_1(x)$ is solved by

$$x_1 = \xi_{1,0}(1-\xi_{1,0}t)^{-1}$$
$$x_2 = -t+\xi_{2,0}$$

from where, for any $\xi_0 \in \mathbb{R}^2$, we have the flow along τ_1 as follows

$$\Phi_{z_1}^{\tau_1}(\xi_0)=\begin{pmatrix}\xi_{1,0}(1-\xi_{1,0}z_1)^{-1}\\-z_1+\xi_{2,0}\end{pmatrix} \tag{6.16}$$

whereas, about $\tau_2(x)$, the equation $\dot{x} = \tau_2(x)$ is solved by

$$x_1 = \cos(t)(1+\xi_{1,0}\exp(-2t))^{-1/2}$$
$$x_2 = -\sin(t)(1+\xi_{2,0}\exp(-2t))^{-1/2}$$

from where, for any $\xi_0 \in \mathbb{R}^2$, we have the flow along τ_2 as

$$\Phi_{z_2}^{\tau_2}(\xi_0) = \begin{pmatrix} \cos(z_2)(1+\xi_{1,0}\exp(-2z_2))^{-1/2} \\ -\sin(z_2)(1+\xi_{2,0}\exp(-2z_2))^{-1/2} \end{pmatrix} \tag{6.17}$$

now, let us define $\Psi(z_1,z_2) \to \Phi_{z1}^{\tau 1} \circ \Phi_{z1}^{\tau 1}(x)$, where "$\circ$" denotes the composition with respect to the argument x. The mapping $\Psi(z_1,z_2)$ corresponds to the integration of both τ_1 and τ_2 vector fields. Thus, one has that the mapping becomes

$$\Psi(z_1 z_2) = \begin{pmatrix} x_1 \\ x_2 \end{pmatrix} = \begin{pmatrix} \cos(z_2)(1-\xi_{1,0}\exp(-2z_2))^{-1/2}(1-\xi_{1,0}z_1)^{-1} \\ -\sin(z_2)(1-\xi_{1,0}\exp(-2z_2))^{-1/2} - z_1 \end{pmatrix} \tag{6.18}$$

note that the map $\Psi(z_1,z_2)$ is not defined on all $\mathbb{R}^2$. In fact, if initial condition $\xi_{1,0}$ is such that $0 < |\xi_{1,0}| < 2((5)\text{-}1)^{1/2}$, then there exists any z_1 such that $1 - \xi_{1,0}z_1 = 0$ even if the distribution $\Delta(x) = \text{span}\,\{\tau_1(x),\tau_2(x)\}$ is involutive on $U = \{x \in \mathbb{R}^2 : x_2 = x_1, 0 < |x_1| < 2((5)\text{-}1)^{1/2}$. That is, $x_1(t)$ is not defined for all initial conditions from the composition of the flows along the vector fields $\tau_1(x)$ and $\tau_2(x)$. However, provided $|1 - \xi_{1,0}z_1| > 0$ and $|1 - \xi_{1,0}\exp(-2z_2)| > 0$, there exists the inverse $z = \Psi^{-1}(x)$ of the mapping $\Psi(z_1,z_2)$ at any x in the neighborhood U^o of $x^o \in U$. Thus, the function $z_2(x_1,x_2)$ is defined and, solves the partial differential equation $\partial z_2/\partial x\tau_1(x) = 0$.

6.2.3 Computing the Synchronization Function

Example 6.18 shows the importance of the Frobenius Theorem for the decompositions along vector fields. However, such an example does not illustrate its usage to compute the synchronization function. In this section we show how the Lie-based geometry of dynamical systems can be exploited in the chaos synchronization context. In seek of simplicity and completeness, the synchronization of nonchaotic systems is illustrated. Then, the synchronization function on chaotic system is computed.

In what follows, let us denote $\tau_1(x_M),\tau_2(x_S)$ and $g(x_S)$ as vector fields such that the synchronization error systems can be defined from the dynamical systems $\dot{x}_M = \tau_1(x_M)$ and $x_S = \tau_2(x_S) + g(x_S)u$, where $x_M \in \mathbb{R}^n$, $x_S \in \mathbb{R}^n$ and $u \in \mathbb{R}$ stands for the control input. By defining $x := x_M - x_S$ as the synchronization error, the synchronization system can be written by $\dot{x} = \tau_1(x_M) - \tau_2(x_S) - g(x_S)u$. Note that $\tau_1(x_M)$ corresponds to the vector fields of the master sysetms whereas $\tau_2(x_S)$ regards the slave system. Then, the synchronization function $y = \lambda(x)$ can be computed from the vector fields $\Delta\tau$: $\mathbb{R}^n \to \mathbb{R}^n$, where $\Delta\tau := \tau_1 - \tau_2$, and $g : \mathbb{R}^n \to \mathbb{R}^n$ by exploiting the Lie algebra. To this end, we use some properties of the Lie derivative, which is related to the inner product $\langle d\lambda(x), \Delta\tau(x)\rangle$ between the vector $d\lambda(x) = (\partial\lambda/\partial x_1, \partial\lambda/\partial x_2, ..., \partial\lambda/\partial x_n)$ and a given vector field $\Delta\tau(x)$ [18], [19].

Now, let $\Delta\tau: \mathbb{R}^n \to \mathbb{R}^n$, $g : \mathbb{R}^n \to \mathbb{R}^n$ be two smooth vector fields. Let us denote $\lambda(x)$ as the real-valued function of $x \in \mathbb{R}^n$ such that its exact derivative is given by $d\lambda(x) = [\partial\lambda(x)/\partial x_{1,\ldots,}\ \partial\lambda(x)/\partial x_n]$. The output $y = \lambda(x)$ of the synchronization error system $x = \Delta\tau(x) + g(x)u$ is a synchronization function if there is a real-valued function $\lambda(x)$ such that: (i) the inner product $\langle d\lambda(x), g(x)\rangle = 0$ for $i = 1,2,\ldots,n-1$ and (ii) $\langle d\lambda(x), ad^i_{\Delta\tau}g(x)\rangle \neq 0$, for any integer (relative degree) $\rho \leq n$ (see Theorem 6.10).

It should be noted that conditions (i) and (ii) imply the existence of a functional relationship, via the real-valued fucntion $\lambda(x)$, between states of the synchronization system (*i.e.*, generalized synchronization). Such conditions are relevant because they signify that the flow of the synchronization-error system can be affected by the scalar function u throughout the vector field $g(x)$ (*i.e.*, since $\langle d\lambda(x), ad^k_{\Delta\tau}\, g(x)\rangle$ denotes the inner product, it relates the orthogonal (tangent) space spaned by the covector field $d\lambda(x)$ and those resulting of the k - th Lie bracket $ad^k_{\Delta\tau}\, g(x)$ (for details see[27] and [28]).

Synchronization Function in Nonchaotic Systems

Let $a \in \mathbb{R}$ and $\rho \in \mathbb{R}^2$ be real positive constants and consider two vector fields τ_1 and τ_2 in $\mathbb{R}^2$ given by

$$\tau_1(x) = \begin{pmatrix} x_2 \\ -ax_1 \end{pmatrix}; \tau_2(x) = \begin{pmatrix} x_2 + x_1(1 - x_2^2 - x_1^2) \\ -x_1 + x_2(1 - x_2^2 - x_1^2) \end{pmatrix}$$

Note that the vector field $\tau_1(x)$ can be related to the harmonic oscillator $\dot{x}_M = \tau_1(x_M)$ (whose fundamental frequency is given by the parameter a, and the amplitude is given by initial conditions) whereas $\tau_2(x)$ is a vector field (which can be related to a dynamical system $x_S = \tau_2(x_S)$ whose attractor is a periodic orbit with fundamental frequency equal to 1; see Example 6.18). Thus, the difference $\Delta\tau(x) := \tau_1(x) - \tau_2(x)$ can be interpreted as the vector field related to the synchronization system between the harmonic and damping second-order oscillators. Now, let us consider the vector field $g(x) = (g_1\ g_2)^T$, where g_1 and g_2 are given constant such that at least one is no null, and a force $u \in \mathbb{R}$ such that the synchronization error system takes the affine form $\dot{x} = \Delta\tau(x) - g(x)u$. In this manner, the problem can be worded as follows: is there any synchronization function $\lambda(x)$ such that the flow along the vector field $\tau_1(x)$ and $\tau_2(x)$ are synchronous? (see Section 6.1). In other words, we shall compute the output $y = \lambda(x)$ such that the synchronization command $g(x)u$ asymptotically steers the trajectories $x(t) := x_M(t) - x_S(t)$ around the origin. That is, the flow along the vector field $\Delta\tau(x)$ converges to zero, to this end, the accesability space must be spaned at the origin. Thus, we need to find the distribution $\Delta = \text{span}\ \{g, ad_{\Delta\tau}g\}$ such that it is involutive, at least, at the origin. By taking the discrepancy vector field, we have that

$$\Delta\tau(x) := \tau_1(x_M) - \tau_2(x_S) = \begin{pmatrix} x_{2M} - x_{2S} - x_{1S}(1 - x_{2S}^2 - x_{1S}^2) \\ -ax_{1M} + x_{1S} - x_{2S}(1 - x_{2S}^2 - x_{1S}^2) \end{pmatrix} \tag{6.19}$$

from where

$$lb_{\Delta\tau}g(x) = [\Delta\tau, g](x)$$
$$= \begin{pmatrix} -2g_1(x_{1M}-x_1)^2 + g_1(1-(x_{2M}-x_2)^2-(x_{1M}-x_1)^2) + g_2(1-2(x_{1M}-x1)(x_{2M}-x_2)) \\ (-1-2(x_{2M}-x_2)(x_{1M}-x_1))g_1 - 2g_2(x_{2M}-x_2)^2 + g_2(1-2(x_{1M}-x1)(x_{2M}-x_2)) \end{pmatrix} \quad (6.20)$$

thus, the distribution has dimension 2 for any x on the set $U = \{x \in \mathbb{R}^2 : -3g_1x_{2M}^2+6g_1x_{2M}x_2-3g_1x_2^2+g_1-g_1x_1^2+2g_1x_{1M}x_1-g_1x_1^2-g_2+2g_2x_{2M}x_{1M}-2g_2x_{2M}x_1-2g_2x_2x_{1M}+2g_2x_2x_1 \neq 0\}$. Hence, Δ = span $\{g, ad_{\Delta\tau}g\}$ is involutive on $U \subset \mathbb{R}^2$. Note that U contains the point (0,0), which gives $U^* = \{x_M \in \mathbb{R}^2 : -3g_1x_{2M}^2 + g_1 - g_2 + 2g_2x_{2M}x_{1M} \neq 0\}$. According to Frobenius Theorem, as a consequence of Δ = span $\{g, lb_{\Delta\tau}g\}$ is involutive, the system $\dot{x} = \Delta\tau(x) - g(x)u$ is integrable. That is, the solution $x(t) = \Phi_t^{\Delta\tau}(x)$ exists for any point x in $U \subset \mathbb{R}^2$. In addition, the integer $\rho = n = 2$.

Now, in order to obtain the synchronization function, we proceed to find the real-valued function $\lambda(x)$ which satisfies conditions (i) and (ii). Thus, we have that the synchronization function $y = \lambda(x)$ should satisfy conditions on Theorem 6.10

$$\langle d\lambda(x), lb_{\Delta\tau}^0 g(x)\rangle = \begin{pmatrix} \frac{\partial\lambda(x)}{\partial x_1} & \frac{\partial\lambda(x)}{\partial x_2} \end{pmatrix} \begin{pmatrix} g_1 \\ g_2 \end{pmatrix} = 0 \quad (6.21)$$

and

$$\langle d\lambda(x), lb_{\Delta\tau}g(x)\rangle = \begin{pmatrix} \frac{\partial\lambda(x)}{\partial x_1} & \frac{\partial\lambda(x)}{\partial x_2} \end{pmatrix}$$
$$\begin{pmatrix} -2g_1(x_{1M}+x_1)^2 + g_1(1-(x_{2M}-x_2)^2-(x_{1M}-x_1)^2) + g_2(1-2(x_{1M}-x_1)(x_{2M}-x_2)) \\ g_1(-1-2(x_{2M}-x_2)(x_{1M}-x_1)) - 2g_2(x_{2M}-x_2)^2 + g_2(1-(x_{2M}-x_2)^2-(x_{1M}-x_1)^2) \end{pmatrix} \neq 0 \quad (6.22)$$

In this manner, for example, if $g = (0\ g_2)^T$, we can consider the real-valued function $\lambda(x) = x_1$ since it satisfies conditions (6.21) and (6.22) if $g_2(1-2x_{2M}x_{1M}) \neq 0$ and can be used as synchronization function. Note that, if synchronization is achieved, the synchronization error $x := x_M - x_S = 0$ then the synchronization function implies that $x_{1,M} = x_{1,S}$; *i.e.*, the phase locking corresponds to a straight line.

Finding the Synchronization Function in Chaotic Systems

Now, let us consider the following vector fields

$$\tau_1(x) = \begin{bmatrix} \pi_1(x_2-x_1) \\ \pi_2x_1 - x_2 - x_1x_3 \\ -\pi_3x_3 + x_1x_2 \end{bmatrix}; \tau_2(x) = \begin{bmatrix} \mu_1(x_2-x_1) \\ \mu_2x_1 - x_2 - x_1x_3 \\ -\mu_3x_3 + x_1x_2 \end{bmatrix}, g = \begin{bmatrix} g_1 \\ g_2 \\ g_3 \end{bmatrix} \quad (6.23)$$

note that the synchronization associated to the vector fields $\tau_1(x)$ and $\tau_2(x)$ can be interpreted as the synchronization of two Lorenz systems with different parameters. Thus, the master system can be written as $\dot{x}_M = \tau_1(x_M)$ and the response system becomes $\dot{x}_S = \tau_2(x_S) + gu$. Thus, the vector field of the synchronization error is given by

$$\Delta\tau(x) := \tau_1(x_M) - \tau_2(x_S) = \begin{bmatrix} \pi_1(x_2 - x_1) - \mu_1(x_2 - x_1) \\ \pi_2 x_1 - \mu_2 x_1 \\ (\mu_3 - \pi_3)x_3 \end{bmatrix} \tag{6.24}$$

the maximum dimension of the spanned space is 3. Thus, by computing the distribution $\Delta(x)$ = span $\{g, ad_{\Delta\tau}g, ad^2_{\Delta\tau}g\}$

$$\Delta(x) = span\left\{\begin{bmatrix} g1 \\ g2 \\ g3 \end{bmatrix}, \begin{bmatrix} (\pi_1 + \mu_1)g_1 - (\pi_1 - \mu_1)g_2 \\ (\mu_2 - \pi_2)g_2 \\ (\pi_3 - \mu_3)g_3 \end{bmatrix}, \begin{bmatrix} (\pi_1 - \mu_1)^2 g_1 - (\pi_1^2 - \mu_1^2)g_2 \\ (\mu_2 - \pi_2)^2 g_2 \\ (\pi_3 - \mu_3)^2 g_3 \end{bmatrix}\right\}$$

by taking the recursive inner product, we have that since the third vector is linearly dependent of the second one, we have that the integer $\rho = 2$ for any difference $\pi_i - \mu_i \neq 0$ and all constant value $g_i \neq 0$, $i = 1,2,3$. Thus conditions (i) and (ii) of Theorem 6.10 becomes

$$\langle d\lambda, ad^k_{\Delta\tau}g\rangle = \begin{cases} 0 & \textit{for any} \quad k = 0 \\ \neq 0 & \textit{for } k = 1 \end{cases} \tag{6.25}$$

in this manner, we have that the vector fields (6.23), which are related to the Lorenz equation, render a maximum value of the integer $\rho = 2$ such that the dimension of the distribution holds at any point x in $\mathbb{R}^3$ for any parameters values satisfying $\pi_i \neq \mu$ $_i$ = 1,2,3 and constants $g_i \neq 0$. In this manner, the function $\lambda(x)$ should satisfy the partial differential equations

$$\begin{aligned} &\langle d\lambda, g\rangle = \frac{\partial\lambda}{\partial x_1}g_1 + \frac{\partial\lambda}{\partial x_2}g_2 + \frac{\partial\lambda}{\partial x_3}g_3 = 0 \\ &\langle d\lambda, lb_{\Delta\tau}g\rangle = ((\pi_1 - \mu_1)g_1 - (\pi_1 + \mu_1)g_2)\frac{\partial\lambda}{\partial x_1} + (\pi_2 + \mu_2)g_1\frac{\partial\lambda}{\partial x_2} + (\pi_3 - \mu_3)g_3 + \frac{\partial\lambda}{\partial x_3} \neq 0 \end{aligned} \tag{6.26}$$

note that $\lambda(x) = g_2x_1 - g_1x_2$ is solution of the equations (6.26), and, as consequence, it corresponds to the synchronization function. Since generalized synchronization is related to the existence of a functional relationship between the states of two systems. In this sense the Lie-based geometry can be exploited to compute the synchronization function $\lambda(x)$ toward a general framework of the synchronization theory, and provides the properties of chaotic systems such that they can be synchronized (*i.e.*, synchronizability of chaotic systems). The novelty in this section consists of the geometrical interpretation of the synchronizability property. Two examples have been presented. The former illustrates the details for computing synchronization function from Lie-based geometry. The later shows the obtainment of the synchronization function on the chaos theory context.

References

[1] Femat, R., Alvarez-Ramirez, J.: Synchronization of a class of strictly different oscillators. Phys. Lett. A 236, 307 (1997)

[2] Ogorzalek, M.J.: Taming chaos-II: Control. IEEE Trans. Circuits and Syst. I 40, 700 (1993)

[3] Brown, R., Kocarev, L.: A unifying framework of chaos synchronization for dynamical systems. Chaos 10, 344–349 (2000)
[4] Boccaletti, S., Pecora, L.M., Pelaez, A.: Unifying framework for chaos synchronization of coupled dynamical systems. Phys. Rev. E 63, 66219 (2001)
[5] Femat, R., Solís-Perales, G.: On the chaos synchronization phenomena. Phys. Lett. A 262, 183 (1999)
[6] Solís-Perales, G.: Sincronizaci6n de Marcha de Polípodos, Ms.Sc. Thesis, UASLP, México (in Spanish) (1999)
[7] Buono, P.L., Golubitsky, M.: Models for central pattern generators for quadruped locomotion I. J. Math. Biol. 42, 291 (2001)
[8] Pasemann, F.: Synchronized chaos and other coherent states for two coupled neurons. Phys. D 128, 236 (1999)
[9] Pérez, G., Cerdeira, H.A.: Extracting mesages masked by chaos. Phys. Rev. Lett. 74, 1970 (1995)
[10] Wu, C.W., Chua, L.O.: A simple way to synchronize chaotic systems with application to secure communication. Int. Jour. Bifur. and Chaos 3, 1619 (1993)
[11] Femat, R., Jauregi-Ortíz, R., Solís-Perales, G.: A chaos-based communication scheme via robust asymptotic feedback. IEEE Trans. Circuits and Syst. I 48, 1161 (2001)
[12] Rosenblum, M.G., Pikovsky, A.S., Kurths, J.: Phase synchronization of chaotic oscillators. Phys. Rev. Lett. 76, 1804 (1996)
[13] Rulkov, N., Sushchik, M.M., Tsimring, L.S., Abarbanel, H.D.I.: Generalized synchronization of chaos in directly coupled chaotic systems. Phys. Rev. E 51, 980 (1995)
[14] Mainieri, R., Rehacek, J.: Projective synchronization in three-dimensional chaotic systems. Phys. Rev. Lett. 82, 3042 (1999)
[15] Udwadia, F.E., Raju, N.: Some global properties of a pair of coupled maps: Quasi-symmetry, periodicity and synchronicity. Physica D 58, 347 (1998)
[16] Josić, K.: Synchronization of chaotic systems and invariant manifolds. Nonlinearity 13, 1321 (2000)
[17] Hermann, R., Krener, A.J.: Nonlinear controlability and observability. IEEE Trans. Automat. Control 22, 728 (1977)
[18] Isidori, A.: Nonlinear Control Systems. Springer, Berlin (1989)
[19] Nijmeijer, H., van der Schaft, A.: Nonlinear Dynamical Control Systems. Springer, New York (1990)
[20] Kailath, T.: Linear Systems. Prentice-Hall, Englewood Clifs (1980)
[21] Holstein-Rathlou, N.H., Yip, K.P., Sosnovtseva, O.V., Mosekilde, E.: Synchronization phenomena in nephron-nephron interaction. Chaos 11, 417 (2001)
[22] Kocarev, L., Parlitz, U.: General approach for chaotic synchronization with applications to communications. Phys. Rev. Letts. 74, 5028 (1995)
[23] Xiaofeng, G., Lai, C.H.: On the synchronization of different chaotic oscillations. Chaos Solitons and Fractals 11, 1231–1235 (2000)
[24] Femat, R., Solís-Perales, G.: Synchronization of chaotic systems with different order. Phys. Rev. E 65, 036226 (2002)
[25] Nijmeijer, H., Marels, M.Y.: An observer looks at synchronization. IEEE Trans. Circuits and Syst. I 44, 307 (1997)
[26] Kocarev, L., Parlitz, U., Hu, B.: Lie derivatives in dynamical systems. Chaos, Solitons and Fractals 9, 1359 (1998)
[27] Solís-Perales, G.: Synchronization of nonlinear systems, Ph.D. Thesis, U.A.S.L.P., México (in spanish) (2002)
[28] Solís-Perales, G., Ayala, V., Klieman, W., Femat, R.: Complete synchronizability of chaotic systems: a geometric approach. Chaos 13, 495 (2002)

Index

Abbreviations

ACA	Actual Control Action
APS	Almost Partial-state Synchronization
CCS	Computed Control Signal
CES	Complete Exact Synchronization
CPS	Complete Practical Synchronization
FFT	Fast Fourier Transform
FPT	Fixed Point Theorem
FLNS	Fully-Linearizable Nonlinear System
GS	Generalized Synchronization
GAS	Globally Asymptotically Stable
HSIC	High-Sensitivity of the Initial Condition
HTS	High T_c Superconductor
IFC	Ideal Feedback Control
OGY	Ott, Grebogy and Yorke
PB	Proportional Band
PII^2	Proportional-Integral-Double Integral
PSD	Power Spectrum Density
PPS	Partial state Practical Synchronization
RAF	Robust asymptotic feedback
RAS	Robust Asymptotic Stabilization
SGT	Small gain theorem
TDC	Time Discrete Coordinate

Lecture Notes in Control and Information Sciences

Edited by M. Thoma, M. Morari

Further volumes of this series can be found on our homepage: springer.com

Vol. 378: Femat R.; Solis-Perales G.
Robust Synchronization of Chaotic Systems via Feedback
199 p. 2008 [978-3-540-69306-2]

Vol. 377: Patan K.
Artificial Neural Networks for the Modelling and Fault Diagnosis of Technical Processes
206 p. 2008 [978-3-540-79871-2]

Vol. 376: Hasegawa Y.
Approximate and Noisy Realization of Discrete-Time Dynamical Systems
245 p. 2008 [978-3-540-79433-2]

Vol. 375: Bartolini G.; Fridman L.; Pisano A.; Usai E. (Eds.)
Modern Sliding Mode Control Theory
465 p. 2008 [978-3-540-79015-0]

Vol. 374: Huang B.; Kadali R.
Dynamic Modeling, Predictive Control and Performance Monitoring
240 p. 2008 [978-1-84800-232-6]

Vol. 373: Wang Q.-G.; Ye Z.; Cai W.-J.; Hang C.-C.
PID Control for Multivariable Processes
264 p. 2008 [978-3-540-78481-4]

Vol. 372: Zhou J.; Wen C.
Adaptive Backstepping Control of Uncertain Systems
241 p. 2008 [978-3-540-77806-6]

Vol. 371: Blondel V.D.; Boyd S.P.; Kimura H. (Eds.)
Recent Advances in Learning and Control
279 p. 2008 [978-1-84800-154-1]

Vol. 370: Lee S.; Suh I.H.; Kim M.S. (Eds.)
Recent Progress in Robotics: Viable Robotic Service to Human
410 p. 2008 [978-3-540-76728-2]

Vol. 369: Hirsch M.J.; Pardalos P.M.; Murphey R.; Grundel D.
Advances in Cooperative Control and Optimization
423 p. 2007 [978-3-540-74354-5]

Vol. 368: Chee F.; Fernando T.
Closed-Loop Control of Blood Glucose
157 p. 2007 [978-3-540-74030-8]

Vol. 367: Turner M.C.; Bates D.G. (Eds.)
Mathematical Methods for Robust and Nonlinear Control
444 p. 2007 [978-1-84800-024-7]

Vol. 366: Bullo F.; Fujimoto K. (Eds.)
Lagrangian and Hamiltonian Methods for Nonlinear Control 2006
398 p. 2007 [978-3-540-73889-3]

Vol. 365: Bates D.; Hagström M. (Eds.)
Nonlinear Analysis and Synthesis Techniques for Aircraft Control
360 p. 2007 [978-3-540-73718-6]

Vol. 364: Chiuso A.; Ferrante A.; Pinzoni S. (Eds.)
Modeling, Estimation and Control
356 p. 2007 [978-3-540-73569-4]

Vol. 363: Besançon G. (Ed.)
Nonlinear Observers and Applications
224 p. 2007 [978-3-540-73502-1]

Vol. 362: Tarn T.-J.; Chen S.-B.; Zhou C. (Eds.)
Robotic Welding, Intelligence and Automation
562 p. 2007 [978-3-540-73373-7]

Vol. 361: Méndez-Acosta H.O.; Femat R.; González-Álvarez V. (Eds.):
Selected Topics in Dynamics and Control of Chemical and Biological Processes
320 p. 2007 [978-3-540-73187-0]

Vol. 360: Kozlowski K. (Ed.)
Robot Motion and Control 2007
452 p. 2007 [978-1-84628-973-6]

Vol. 359: Christophersen F.J.
Optimal Control of Constrained Piecewise Affine Systems
190 p. 2007 [978-3-540-72700-2]

Vol. 358: Findeisen R.; Allgöwer F.; Biegler L.T. (Eds.): Assessment and Future Directions of Nonlinear Model Predictive Control
642 p. 2007 [978-3-540-72698-2]

Vol. 357: Queinnec I.; Tarbouriech S.; Garcia G.; Niculescu S.-I. (Eds.):
Biology and Control Theory: Current Challenges
589 p. 2007 [978-3-540-71987-8]

Vol. 356: Karatkevich A.:
Dynamic Analysis of Petri Net-Based Discrete Systems
166 p. 2007 [978-3-540-71464-4]

Vol. 355: Zhang H.; Xie L.:
Control and Estimation of Systems with Input/Output Delays
213 p. 2007 [978-3-540-71118-6]

Vol. 354: Witczak M.:
Modelling and Estimation Strategies for Fault Diagnosis of Non-Linear Systems
215 p. 2007 [978-3-540-71114-8]

Vol. 353: Bonivento C.; Isidori A.; Marconi L.; Rossi C. (Eds.)
Advances in Control Theory and Applications
305 p. 2007 [978-3-540-70700-4]

Vol. 352: Chiasson, J.; Loiseau, J.J. (Eds.)
Applications of Time Delay Systems
358 p. 2007 [978-3-540-49555-0]

Vol. 351: Lin, C.; Wang, Q.-G.; Lee, T.H., He, Y.
LMI Approach to Analysis and Control of Takagi-Sugeno Fuzzy Systems with Time Delay
204 p. 2007 [978-3-540-49552-9]

Vol. 350: Bandyopadhyay, B.; Manjunath, T.C.; Umapathy, M.
Modeling, Control and Implementation of Smart Structures 250 p. 2007 [978-3-540-48393-9]

Vol. 349: Rogers, E.T.A.; Galkowski, K.; Owens, D.H.
Control Systems Theory and Applications for Linear Repetitive Processes
482 p. 2007 [978-3-540-42663-9]

Vol. 347: Assawinchaichote, W.; Nguang, K.S.; Shi P.
Fuzzy Control and Filter Design for Uncertain Fuzzy Systems
188 p. 2006 [978-3-540-37011-6]

Vol. 346: Tarbouriech, S.; Garcia, G.; Glattfelder, A.H. (Eds.)
Advanced Strategies in Control Systems with Input and Output Constraints
480 p. 2006 [978-3-540-37009-3]

Vol. 345: Huang, D.-S.; Li, K.; Irwin, G.W. (Eds.)
Intelligent Computing in Signal Processing and Pattern Recognition
1179 p. 2006 [978-3-540-37257-8]

Vol. 344: Huang, D.-S.; Li, K.; Irwin, G.W. (Eds.)
Intelligent Control and Automation
1121 p. 2006 [978-3-540-37255-4]

Vol. 341: Commault, C.; Marchand, N. (Eds.)
Positive Systems
448 p. 2006 [978-3-540-34771-2]

Vol. 340: Diehl, M.; Mombaur, K. (Eds.)
Fast Motions in Biomechanics and Robotics
500 p. 2006 [978-3-540-36118-3]

Vol. 339: Alamir, M.
Stabilization of Nonlinear Systems Using Receding-horizon Control Schemes
325 p. 2006 [978-1-84628-470-0]

Vol. 338: Tokarzewski, J.
Finite Zeros in Discrete Time Control Systems
325 p. 2006 [978-3-540-33464-4]

Vol. 337: Blom, H.; Lygeros, J. (Eds.)
Stochastic Hybrid Systems
395 p. 2006 [978-3-540-33466-8]

Vol. 336: Pettersen, K.Y.; Gravdahl, J.T.; Nijmeijer, H. (Eds.)
Group Coordination and Cooperative Control
310 p. 2006 [978-3-540-33468-2]

Vol. 335: Kozłowski, K. (Ed.)
Robot Motion and Control
424 p. 2006 [978-1-84628-404-5]

Vol. 334: Edwards, C.; Fossas Colet, E.; Fridman, L. (Eds.)
Advances in Variable Structure and Sliding Mode Control
504 p. 2006 [978-3-540-32800-1]

Vol. 333: Banavar, R.N.; Sankaranarayanan, V.
Switched Finite Time Control of a Class of Underactuated Systems
99 p. 2006 [978-3-540-32799-8]

Vol. 332: Xu, S.; Lam, J.
Robust Control and Filtering of Singular Systems
234 p. 2006 [978-3-540-32797-4]

Vol. 331: Antsaklis, P.J.; Tabuada, P. (Eds.)
Networked Embedded Sensing and Control
367 p. 2006 [978-3-540-32794-3]

Vol. 330: Koumoutsakos, P.; Mezic, I. (Eds.)
Control of Fluid Flow
200 p. 2006 [978-3-540-25140-8]

Vol. 329: Francis, B.A.; Smith, M.C.; Willems, J.C. (Eds.)
Control of Uncertain Systems: Modelling, Approximation, and Design
429 p. 2006 [978-3-540-31754-8]

Vol. 328: Loría, A.; Lamnabhi-Lagarrigue, F.; Panteley, E. (Eds.)
Advanced Topics in Control Systems Theory
305 p. 2006 [978-1-84628-313-0]

Vol. 327: Fournier, J.-D.; Grimm, J.; Leblond, J.; Partington, J.R. (Eds.)
Harmonic Analysis and Rational Approximation
301 p. 2006 [978-3-540-30922-2]

Vol. 326: Wang, H.-S.; Yung, C.-F.; Chang, F.-R.
H_∞ Control for Nonlinear Descriptor Systems
164 p. 2006 [978-1-84628-289-8]

Vol. 325: Amato, F.
Robust Control of Linear Systems Subject to Uncertain Time-Varying Parameters
180 p. 2006 [978-3-540-23950-5]